RENEWABLE ENERGY: RESEARCH, DEVELOPMENT AND POLICIES

THE CONSUMER COMPONENT IN RENEWABLE ENERGY USE

SELECT RESEARCH

RENEWABLE ENERGY: RESEARCH, DEVELOPMENT AND POLICIES

Additional books in this series can be found on Nova's website under the Series tab.

Additional E-books in this series can be found on Nova's website under the E-book tab.

ENERGY SCIENCE, ENGINEERING AND TECHNOLOGY

Additional books in this series can be found on Nova's website under the Series tab.

Additional E-books in this series can be found on Nova's website under the E-book tab.

RENEWABLE ENERGY: RESEARCH, DEVELOPMENT AND POLICIES

THE CONSUMER COMPONENT IN RENEWABLE ENERGY USE

SELECT RESEARCH

GEORGIA M. LEE
AND
RUSS L. COOPER
EDITORS

Nova Science Publishers, Inc.
New York

For permission to use material from this book please contact us:
Telephone 631-231-7269; Fax 631-231-8175
Web Site: http://www.novapublishers.com

Library of Congress Cataloging-in-Publication Data

The consumer component in renewable energy use : select research / editors, Georgia M. Lee and Russ L. Cooper.
p. cm.
Includes index.
ISBN 978-1-62081-425-3 (hardcover)
1. Renewable energy sources--Public opinion--United States 2. Energy consumption--United States. 3. Consumers' preferences--United States. I. Lee, Georgia M. II. Cooper, Russ L.
TJ807.9.U6C66 2012
333.79'4--dc23
2012008879

Published by Nova Science Publishers, Inc. †New York

CONTENTS

PREFACE

Renewable energy is available to consumers in three forms: (1) utility green pricing programs offered in regulated states, (2) competitive retail electricity products in deregulated or competitive markets, and (3) renewable energy certificates (RECs) purchased in over-the-counter transactions or through independent REC marketers. The objective of this book is to explore trends in consumer interest and awareness of renewable energy options through analysis of historical and regional survey data. This book documents green power marketing activities and trends in the United States and discusses the social acceptance of utility-scale renewable energy projects.

Chapter 1 - The data in this report are taken from Natural Marketing Institute's (NMI) Lifestyles of Health and Sustainability (LOHAS) Consumer Trends Database®. Created in 2002, the syndicated consumer database contains responses from 2,000 to 4,000 nationally representative US adults (meaning the demographics of the sample are consistent with US Census findings) each year.

A sample of 2,000 has a confidence interval of +/-2.2 percentage points at the 95% confidence level, which decreases to +/- 1.2 percentage points with 4,000 respondents. NMI used the database to analyze consumer attitudes and behavior related to renewable energy and to update previously conducted related research. Specifically, this report will explore consumer awareness, concerns, perceived benefits, knowledge of purchase options, and usage of renewable energy as well as provide regional comparisons and trends over time.

Based on this analysis, NMI found the following:

- The majority of consumers (80%) indicated that they care about the use of renewable energy. However, concern has diminished slightly over time, which is consistent with other broad environmental consumer attitudes (such as concerns related to environmental protection or sustainable agriculture).
- Consumers primarily associate renewable energy with environmental benefits, despite the other potential benefits renewable energy has to offer and the recent efforts to broaden its appeal.
- Consumer awareness of renewable energy purchase options remains relatively low, with approximately one in six consumers aware of the green power options provided by their electric suppliers, although approximately half of consumers have options available to them.

- Consumers are more price sensitive for renewable energy than in the past, mirroring an increased price sensitivity NMI has observed across the green consumer landscape.
- Despite the common perception, there are few differences in consumer attitudes across regions. However, consumers in the West are more aware of renewable energy terminology, such as renewable power and carbon footprint. In addition, Western consumers, compared to Midwestern consumers, are more aware of their purchase options, are less price sensitive, and are more likely to have already purchased at least some type of renewable energy.
- Seven percent of the adult population reports buying at least some renewable energy for their home. According to similar research, this is a significantly lower percentage than the proportion of the population that cares about renewable energy and a higher percentage than penetration rates reported by utilities and marketers that offer renewable energy options to consumers.

Opportunities exist for continued market growth in renewable energy as indicated by the differential between concern and usage and declining premiums.

However, the modest awareness levels of renewable energy options are a challenge.

Chapter 2 - This report documents green power marketing activities and trends in the United States. Aggregate green power sales data for all voluntary purchase markets across the United States are presented for 2009. The data presented in this report are based primarily on figures provided to NREL by utilities and independent renewable energy marketers. Because data cannot be obtained from all market participants, the estimates presented here likely represent an underestimate of the market size. Key trends identified in this year's report include:

- In 2009, total retail sales of renewable energy in voluntary markets exceeded 30 million MWh, an increase of 17% from 2008. The increase was dominated by renewable energy certificate (REC) sales, primarily to nonresidential consumers, which increased by about 20% from 2008 (see Figure ES-1). REC markets now represent 62% of total voluntary green power market sales.
- Utility green pricing programs in regulated electricity markets continued to grow on a sales basis but at a slower rate than in previous years, with sales volume increasing by about 7% in 2009. A relatively small number of utility programs continued to dominate sales and customer numbers. Utility premiums for green pricing continued to fall due in part to the increased cost competitiveness of renewable with conventional generation.
- Wind energy provided 73.7% of total green power sales volume, followed by biomass energy sources including landfill gas (10.0%), hydropower (9.9%), geothermal (0.2%), and solar (0.1%), with the remainder unknown (5.9%).
- Overall, the total number of customers purchasing green power increased by 44% in 2009, which is a higher rate than in previous years and with gains coming primarily from a competitive offering in Texas introduced in 2009. Utility green pricing program participants remained essentially flat in aggregate from 2007 to 2009, with some programs continuing to report customer losses in 2009.

In 2009, nearly 340,000 metric tons of avoided CO2e from renewable energy facilities were marketed as offsets, an increase of approximately 39% from 2008. This is the equivalent of about 485,000 MWh of renewable energy generation.

Chapter 3 - The study "Social, Economic, and Policy Factors Affecting the Siting of a 100-MW Concentrating Solar Power (CSP) Facility in Colorado's San Luis Valley" was a joint effort of the University of Colorado (CU) and the National Renewable Energy Laboratory (NREL). The CU study, conducted at the Institute of Behavioral Science, was funded by NREL and the Energy Initiative—now the Renewable and Sustainable Energy Institute (a joint institute of CU and NREL).

The period of performance for the study was originally November 2006 through February 2009. The NREL portion of the study was funded by contract to co-principal investigator Barbara Farhar, the first author of this report. The period of performance for the contract was October 9, 2008 through March 31, 2010.

A Research Advisory Committee provided guidance on key questions and groups to include in the study. Committee members were: Mark Mehos, NREL; Fred Morse, Chairman, Utility-Scale Solar Power Division, Solar Energy Industries Association; Mike Wisdom, San Luis Valley Regional Development Group; Mike Bowman, 25 x '25; and Marty Smith, Xcel Energy.

Part of this report is based on interviews with representatives of 25 key stakeholder groups within and outside the San Luis Valley. The CU principal investigator was Kathleen J. Tierney, Ph.D., and Lori M. Hunter, Ph.D., was also a co-principal investigator. Tracy M. Kirkland, a doctoral candidate in the Department of Sociology, provided graduate research assistance on the joint project. Other research assistance was provided to Dr. Farhar by Gregg Eisenberg (Iron Mountain Consulting) who did the primary work on the transmission issues after the interviewing ended, and Linda Giudice (graduate student, University of Colorado at Boulder) who did the primary work on the Bureau of Land Management Programmatic Environmental Impact Statement process. Therese Edwards and Teresa Foster provided research assistance on coding. Sarah Leshan provided editorial support. Many helpful suggestions on the manuscript were made by reviewers Adam Reed, Center for Energy and Environmental Security, University of Colorado Law School; Timothy Schoechle, Consultant; Fred Morse; and Bill Mansheim, Adams State College. Any errors are the responsibility of the authors.

In: The Consumer Component in Renewable Energy Use
Editors: Georgia M. Lee and Russ L. Cooper ISBN 978-1-62081-425-3

Chapter 1

CONSUMER ATTITUDES ABOUT RENEWABLE ENERGY: TRENDS AND REGIONAL DIFFERENCES*

National Renewable Energy Laboratory

EXECUTIVE SUMMARY

Significant Results, Trends, and Regional Differences

The data in this report are taken from Natural Marketing Institute's (NMI) Lifestyles of Health and Sustainability (LOHAS) Consumer Trends Database®. Created in 2002, the syndicated consumer database contains responses from 2,000 to 4,000 nationally representative US adults (meaning the demographics of the sample are consistent with US Census findings) each year.

A sample of 2,000 has a confidence interval of +/-2.2 percentage points at the 95% confidence level, which decreases to +/- 1.2 percentage points with 4,000 respondents. NMI used the database to analyze consumer attitudes and behavior related to renewable energy and to update previously conducted related research. Specifically, this report will explore consumer awareness, concerns, perceived benefits, knowledge of purchase options, and usage of renewable energy as well as provide regional comparisons and trends over time.

Based on this analysis, NMI found the following:

- The majority of consumers (80%) indicated that they care about the use of renewable energy. However, concern has diminished slightly over time, which is consistent with other broad environmental consumer attitudes (such as concerns related to environmental protection or sustainable agriculture).

* This is an edited, reformatted and augmented version of a National Renewable Energy Laboratory, Contract No. DE-AC36-08GO28308 publication, Prepared under Subcontract No. AGG-1-11880-01, dated April 2011.

- Consumers primarily associate renewable energy with environmental benefits, despite the other potential benefits renewable energy has to offer and the recent efforts to broaden its appeal.
- Consumer awareness of renewable energy purchase options remains relatively low, with approximately one in six consumers aware of the green power options provided by their electric suppliers, although approximately half of consumers have options available to them.
- Consumers are more price sensitive for renewable energy than in the past, mirroring an increased price sensitivity NMI has observed across the green consumer landscape.
- Despite the common perception, there are few differences in consumer attitudes across regions. However, consumers in the West are more aware of renewable energy terminology, such as renewable power and carbon footprint. In addition, Western consumers, compared to Midwestern consumers, are more aware of their purchase options, are less price sensitive, and are more likely to have already purchased at least some type of renewable energy.
- Seven percent of the adult population reports buying at least some renewable energy for their home. According to similar research, this is a significantly lower percentage than the proportion of the population that cares about renewable energy and a higher percentage than penetration rates reported by utilities and marketers that offer renewable energy options to consumers.

Opportunities exist for continued market growth in renewable energy as indicated by the differential between concern and usage and declining premiums.

However, the modest awareness levels of renewable energy options are a challenge.

1. Introduction

1.1. Project Background and Objective

The voluntary renewable energy market has grown considerably since its inception in the 1990s. In 2009, more than 30 million MWh of renewable energy were sold in the voluntary market, a nearly four-fold increase compared to 2005 and almost equal to the amount of new renewables required to meet 2009 state renewable portfolio standards.

Renewable energy is available to consumers in three forms: (1) utility green pricing programs offered in regulated states, (2) competitive retail electricity products in deregulated or competitive markets, and (3) renewable energy certificates (RECs) purchased in over-the-counter transactions or through independent REC marketers.

Customer participation in utility green pricing programs and competitive market programs has increased over the years, though penetration rates remain low. More than 1.4 million consumers purchase green power through their utility or retail electricity provider or in the REC market, with the vast majority (99%) of consumers purchasing through their utility or retail electricity provider.

However, in 2009, the average participation rate among utility green pricing programs was 2.0%, while penetration rates in competitive markets ranged from 1.7% to 2.5%. The top 10 programs in 2009 had participation rates ranging from 5.1% to 20.8%. (Bird and Sumner 2010)

The objective of this report is to explore trends in consumer interest and awareness of renewable energy options through analysis of historical and regional survey data. Specifically, the report provides a historical and regional analysis of consumer responses to questions that indicate how much people care about renewable energy sources, how aware they are of options to buy renewable energy, their willingness to pay for renewable energy, and their opinions on the benefits of renewable energy. Only the renewable energy data are summarized here, provided by Natural Marketing Institute's (NMI's) Lifestyles of Health and Sustainability (LOHAS) Consumer Trends Database, a survey instrument conducted annually since 2002 that covers a broad range of environmental issues. This report also provides an update to previously conducted research on consumer attitudes regarding renewable energy. Earlier consumer market research revealed consumer interest in supporting renewable energy. For example, utility market research on consumer preferences found that 52%–95% of residential consumers were willing to pay at least a modest amount ($3– $10) for renewable energy on a monthly basis (Farhar 1999). Deliberative polling of Texas electricity consumers also found that the median willingness to pay among consumers in various utility territories ranged from $1.50 to $6.50 per month (Lehr et al. 2003). In addition, a 2007 Opinion Research Corporation poll found similar results to 2007 NMI data regarding familiarity with the term renewable energy, awareness of purchasing options, and purchase rates.[1]

The updated data on consumer attitudes and awareness presented here are useful for utilities or marketers engaged in the voluntary renewable energy market. The data on trends and regional differences in attitudes towards renewable energy may also be useful for policymakers, stakeholders, and the renewable energy industry as a whole.

1.2. US LOHAS Consumer Trends Database Research Methodology

1.2.1. Survey Scope

LOHAS describes the consumers and organizations that put health and sustainability first and foremost in their lives. NMI first designed the LOHAS Consumer Trends Database in 2002 to measure and describe the marketplace for LOHAS products and services within the total US population, the consumers who use them, consumer expectations of corporate behavior, and attitudes toward environmental and social issues. Particular attention is paid to consumer attitudes, behaviors, psychographics, lifestyle activities, and product and service usage patterns in order to provide the information needed to capitalize on growing sustainability and corporate responsibility initiatives. Renewable energy has been included since the beginning of the research.

1.2.2. Data Collection

Undertaken by NMI on behalf of its sponsor clients, the 2010 research was conducted in July via an online survey of 4,000 US general population adults. Survey participants included those who had opted in to a panel as well as a demographically balanced, random subset of the panel. As respondents completed the survey, additional sample data were sent to provide balance demographics for any groups that were underrepresented at that point. To correct for any minor variations in the sample's demographic characteristics and the population as a whole, the data have been post-weighted to match multiple US Census demographic measures. No individual respondent's weighting factor is greater than three or less than 0.3. The study utilized a leading online research firm and was designed, managed, and analyzed by NMI. The text of the questions used in this report is included in Appendix A.

1.2.3. Statistical Significance

The results of this survey are nationally projectable to the US adult population and statistically valid at the 95% confidence level +/- 1.2%. Throughout this report, capital letters indicate statistically significant differences between regions at the 95% confidence level using *t*-tests. The *t*-tests reflect the probability that the difference in two numbers being compared is not due to sampling error. In regional charts and graphs, each region is assigned a letter as a short-hand code. A significant regional difference is noted on the region with the higher value by referencing the capital letter of the region with the lower value.

1.2.4. Geographic Division

This report includes analyses by geographic region. The regions are defined in Table 1 and are consistent with industry and market research standards.

Table 1. Report Categories by Geographic Region

	Northeast		Midwest		South		West
•	Connecticut	•	Illinois	•	Alabama	•	Alaska
•	Delaware	•	Indiana	•	Arkansas	•	Arizona
•	District of Columbia	•	Iowa	•	Florida	•	California
•	Maine	•	Kansas	•	Georgia	•	Colorado
•	Massachusetts	•	Michigan	•	Kentucky	•	Hawaii
•	New Hampshire	•	Minnesota	•	Louisiana	•	Idaho
•	New Jersey	•	Missouri	•	Maryland	•	Montana
•	New York	•	Nebraska	•	Mississippi	•	Nevada
•	Pennsylvania	•	North Dakota	•	North Carolina	•	New Mexico
•	Rhode Island	•	Ohio	•	Oklahoma	•	Oregon
•	Vermont	•	South Dakota	•	South Carolina	•	Utah
•	West Virginia	•	Wisconsin	•	Tennessee	•	Washington
				•	Texas	•	Wyoming
				•	Virginia		

2. CONSUMER PERCEPTIONS OF RENEWABLE ENERGY

2.1. Consumer Awareness of Renewable-energy-related Terminology, Trended

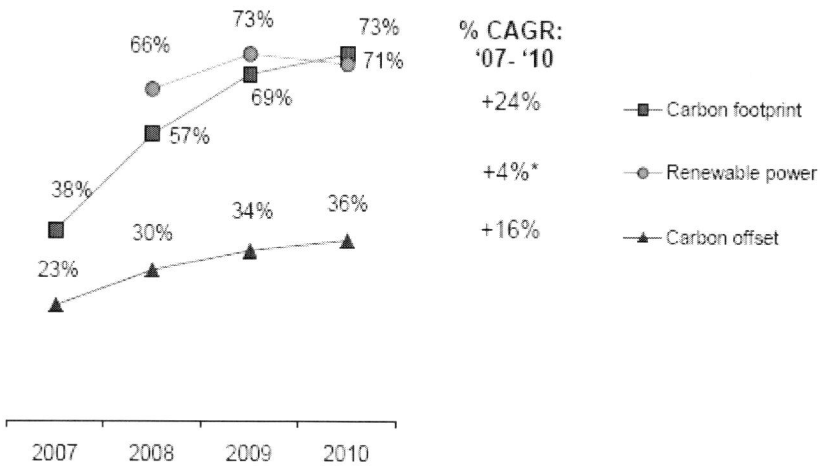

* % Compound annual growth rate (CAGR) 2008–2010[2].
Source: 2010 LOHAS Consumer Trends Database.

Figure 1. In Q.14, percent of the general population indicating awareness of renewable-energy-related terminology.

Generating market interest in renewable energy starts with raising awareness of commonly used terms. Since NMI began measuring awareness of renewable power in 2008, at least two-thirds of consumers have been aware of the term, with a statistically higher number in 2010 (as shown in Figure 1). In the same timeframe, consumer awareness of carbon-related terminology has increased more rapidly. The awareness of the term carbon footprint has grown particularly quickly, up 24% annually since 2007, with most of the growth between 2007 and 2008, when media coverage of this issue was more prevalent, especially in the Northeast and West where there are established carbon regulations (i.e., RGGI in the East and Western Climate Initiative in California). As newer terms, and those commonly used by the media, many consumers have latched onto them.

The term carbon offset has not seen the same dramatic increase in awareness as carbon footprint and is still below the 2007 carbon footprint awareness level. While the two terms may be clearly linked within the business community, the same cannot be said of consumers. While carbon footprints are commonly discussed in consumer media, specifically when talking about how to reduce a carbon footprint, reports often refer to energy efficiency and conservation, use of renewable energy, and carbon offsets, among other approaches. Since carbon offsets are just one of the approaches to manage a carbon footprint, it could explain why awareness of this term is lower.

Interestingly, the term carbon footprint has overtaken the term renewable power in awareness. Since the term renewable power has been used for decades, it is remarkable that carbon footprint has eclipsed it after just a few years. While more consumers may be aware of specific types of renewable energy, such as solar and wind (85% of consumers were aware of wind energy in 2009, for instance), perhaps the umbrella term is not as well known. Therefore, communicators in this field need to be mindful of consumer familiarity with these terms in their publications.

As a comparison, 85% of consumers are aware of the term "biodegradable" (the highest scoring term measured) and 60% are aware of the term "sustainability." Awareness of both renewable power and carbon footprint, therefore, is comparable to other green industry terms. Of course, awareness is *not* the same as understanding. While awareness is a starting point, presumably fewer understand the terms. Ensuring accurate understanding is also important to generating interest in the marketplace.

2.2. Consumer Awareness of Renewable-energy-related Terminology, by Region

Table 2. In Q.14, Percent of the General Population, By Region, Indicating Awareness of Renewable-energy-related Terminology

	Northeast (A)	Midwest (B)	South (C)	West (D)
Carbon footprint	74%	71%	70%	78% BC
Renewable power	68%	71%	70%	78% ABC
Carbon offset	33%	34%	34%	44% ABC

Note: Capital letters indicate significant differences between regions at the 95% confidence level. For example, we are 95% confident that consumers in the West are more aware of "carbon footprint" than consumers in the Midwest (B) and South (C).

Source: 2010 LOHAS Consumer Trends Database.

As shown in Table 2, consumers on the West Coast are significantly more aware of these three terms than consumers in other regions who have statistically equivalent awareness. West Coast residents are often thought to be more attuned to environmental issues than consumers elsewhere in the country. However, even in the West, nearly one-quarter of the population is unaware of the term renewable power.

2.3. Consumer Caring about Using Renewable Energy, Trended

According to Figure 2, a strong majority of consumers care about renewable energy, ranging from about 80% to 90% over the survey period. Slightly fewer consumers care about using renewable energy now than they did in 2002, down 1% annually over the nine-year horizon. This decline has come primarily from the people who agree completely (down 3% annually), meaning that there is an absolute decline and a decline in intensity.

Note that more consumers report concern about use of renewable energy than are aware of the term renewable power. There are a few possible explanations for this apparent disconnect. First, the terms renewable energy and renewable power have subtle but important differences.[3] Second, the question format may affect response; for example, the awareness question asked respondents to check all they were aware of, whereas this question uses a five-point Likert scale format.

A question using with a Likert scale asks respondents to answer using a scale (in this case, from "agree completely" to "disagree completely"), which is more appropriate for a question about concern or caring and also elicits a more thoughtful response.

As shown in Figure 3, concerns about other broad-based environmental issues have also experienced this decline. While the decline may seem ironic during a decade that saw significant growth in renewable energy development and considerable expansion in the number of green products on the market, the trends may actually be related.

The increased development and availability of renewable energy may indicate to consumers that the associated environmental problems have been solved, or are at least being addressed, which may allay their concerns. However, the data between 2009 and 2010 are stable, which could signal the end of the decline.

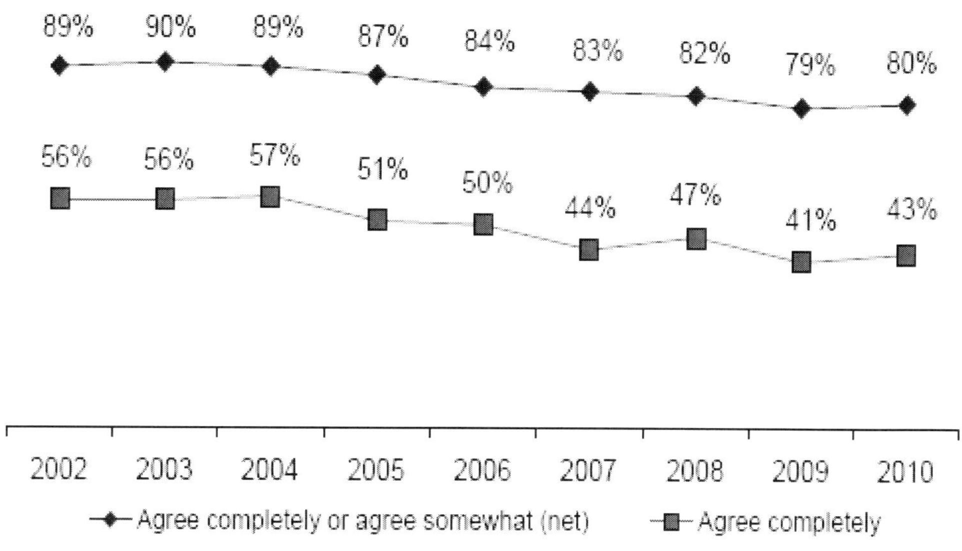

Source: 2010 LOHAS Consumer Trends Database.

Figure 2. In Q.9, percent of the general population indicating that they agree completely or agree completely/somewhat with the statement, "I care about use of renewable energy sources".

2.4. Consumer Caring about Using Renewable Energy in the Context of Other Environmental Attitudes

As mentioned above, and displayed in Figure 3, other broad-based environmental attitudes have declined since NMI started measuring these attitudes in 2002, and the change in renewable energy is consistent with the other changes.

In addition to product proliferation potentially affecting consumer concerns, the overall level of discourse may be leading consumers to think that there is less reason for concern. Environmental issues are in the mainstream news frequently, and it would be reasonable for consumers to believe that with all of the discussion that conditions are actually improving, or are at least being addressed.

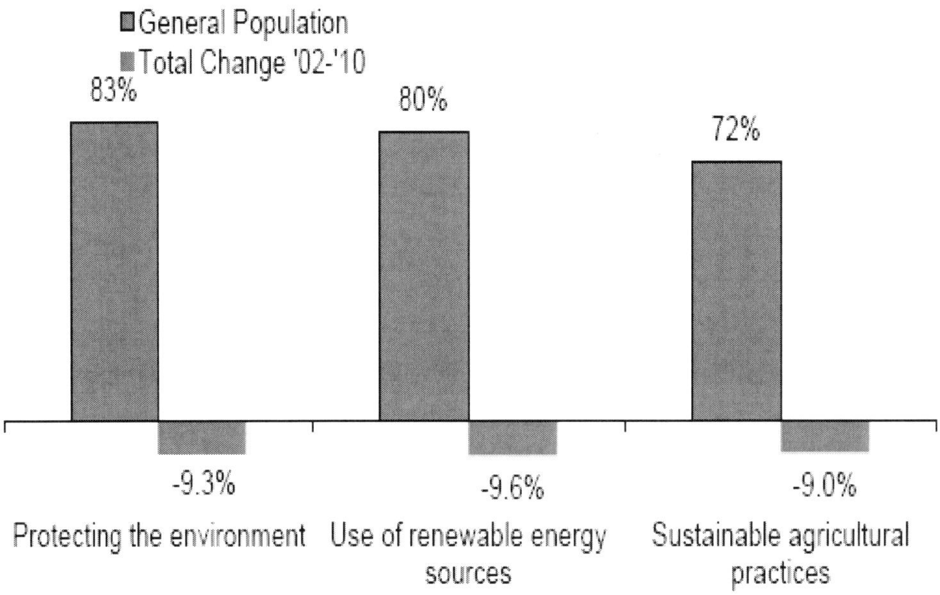

Source: 2010 LOHAS Consumer Trends Database.

Figure 3. In Q.9, percent CAGR among the general population indicating that they agree completely/somewhat with the statement, "I care about..."

As consumer concerns appear to be diminishing, consumers are also doing more than ever to help the environment. More consumers are buying green products, for example, usage of compact fluorescent lighting (CFL) is up 15% annually since 2003 and the purchase of organic foods is up 6% annually in the same time frame. In addition, more than three times as many consumers take their own reusable bags to the store and almost twice as many say they avoid product brands with practices they do not agree with, among dozens of other behavioral changes. It is somewhat unusual for attitudes to decline while behaviors grow; however, many of these behaviors have become less burdensome for consumers, which is probably contributing to their rise.

2.5. Consumer Caring about Using Renewable Energy, by Region

As shown in Figure 4, unlike awareness of renewable energy terminology, wherein consumers in the West had the highest awareness, consumers in the Northeast are slightly more likely to care about the use of renewable energy sources, particularly in comparison to consumers in the South. The difference between consumers in the Northeast and consumers in the South is significant at the 95% confidence level. Even with the Northeastern region's slight lead over the rest of the country, the concern for renewable energy is nearly equal across regions.

This question asks consumers to pick a single, primary benefit of renewable power, though all of the choices could be seen as benefits. A stable plurality of consumers associate renewable power primarily with environmental benefits, as indicated in Figure 5.

Other benefits are much less likely to be cited as the primary benefit, even those benefits that have recently been touted heavily, namely the economic benefits.

The economic benefits are actually on a downward decline, by almost 50% from their 2007 level (a statistically significant decline). Following the trend of earlier data, few differences are evident by region, though the South is slightly more likely to associate renewable power with economic benefits than the Midwest.

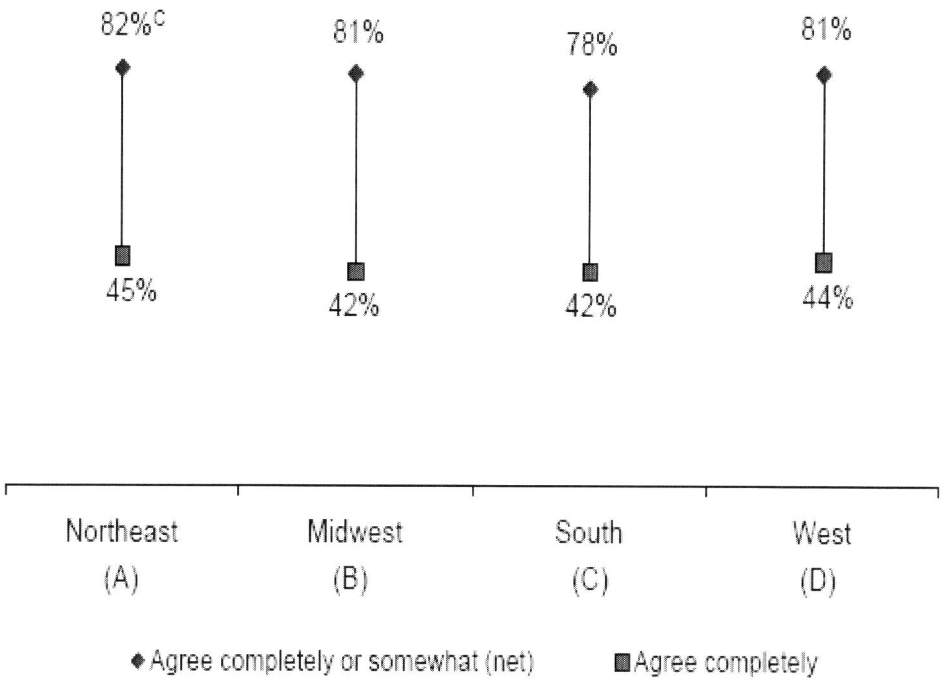

Capital letters indicate significant differences between regions at the 95% confidence level.
Source: 2010 LOHAS Consumer Trends Database.

Figure 4. In Q.9, percent of the general population, by region, indicating that they agree completely or agree completely/somewhatwith the statement, "I care about use of renewable energy sources".

One strategy to grow renewable energy usage beyond its current level (7%) would be to better promote the health and economic benefits, since this would widen the target audience, and it is easier to reposition a product to be relevant to consumers than it is to convince them of its relevance.

As Figure 5 shows, the messaging related to health and economic benefits has not yet been effective; changing these perceptions would require very persuasive educational campaigns.

In addition, to achieve further market growth, research is needed to better understand whether these benefits are not believable, not as important, or if there are other barriers to consumer acceptance.

3. CONSUMER PURCHASING OF RENEWABLE ENERGY

3.1. Consumer Awareness of Renewable Power Purchase Options, Trended

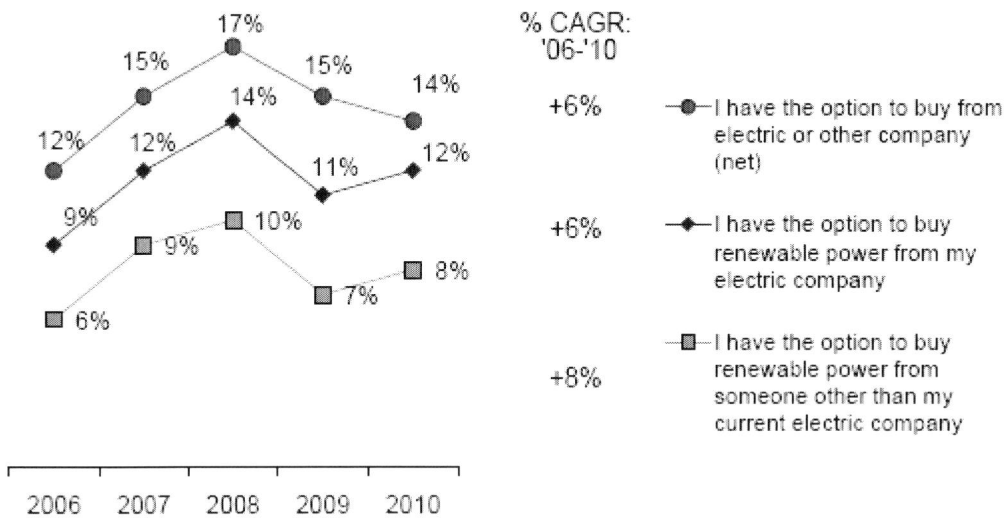

Source: 2010 LOHAS Consumer Trends Database.

Figure 6. In Q.122, percent of the general population indicating their choice of renewable power purchase options.

As depicted in Figure 6, approximately one in six adults know that they have the option to buy renewable power, either from their electric company or another provider. Consumers are much more likely to know they can get renewable power from their current electric provider since they are existing customers and receive regular communications with their monthly bills. Compared to the findings reported by NREL (Bird and Sumner 2010), the 14% of consumers who know they can buy renewable power through their utility or another company is considerably less than those who actually can (more than 50% of consumers have the option to purchase some type of green power product directly from a retail electricity provider, while all consumers have the option to buy RECs.

Compared to 2006, there has been little change in consumer awareness of purchase options, despite a peak in 2008 that did not last. Providers may have pulled back on promotions, assuming that in the height of the recession consumers had other priorities for their discretionary spending. There appears to be a bit of a correction in that, however, between 2009 and 2010.

Because electricity is such a low involvement category, meaning that consumers do not deliberate this decision extensively (if at all) or experience any consequences of their decision, these low figures are not surprising.[4] Nevertheless, there is a wide gulf between the number of consumers who reportedly care about renewable energy and those who know they can buy it (let alone those who actually do, which will be covered later in this report). If voluntary purchases are to be important to growing the renewable energy market, consumers obviously need to know they have the option of putting their money where their values are.

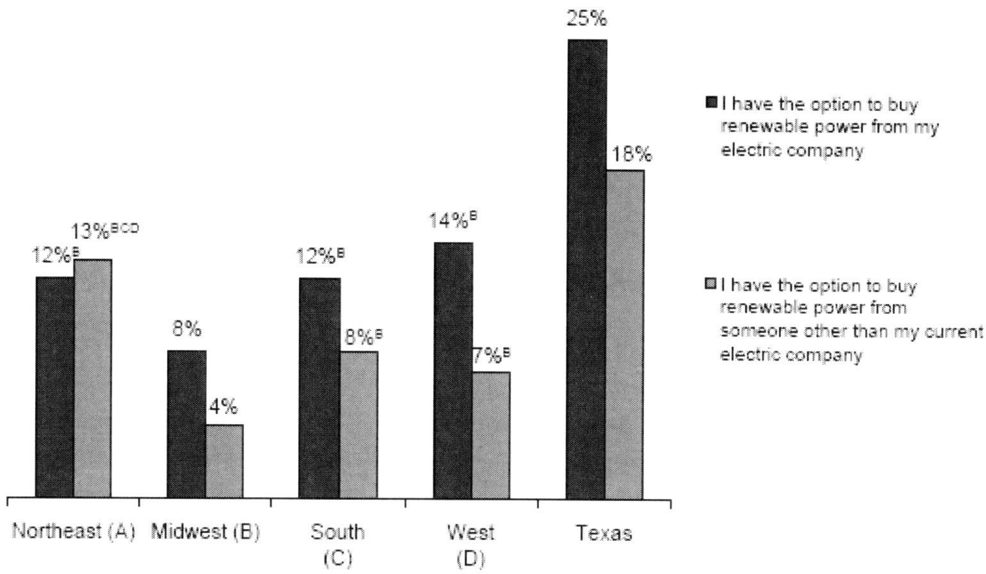

Capital letters indicate significant differences between regions at the 95% confidence level .
Source: 2010 LOHAS Consumer Trends Database.

Figure 7. In Q.122, percent of the general population, by region, indicating their renewable power purchase options.

As shown in Figure 7, there are regional variations in consumer awareness of renewable power purchase options—above and beyond most other regional differences examined in this report. Midwestern consumers are least likely to know of their options, while those on the coasts are more aware. Northeastern consumers, where retail competition exists and several alternative providers operate, are relatively more aware of both options, and the most aware of buying from an alternate supplier. In fact, this is the only region where equal numbers of consumers recognize that they have the option of buying from either type of supplier.

The most striking difference is in Texas, where one-quarter of consumers are aware of renewable power purchase options. The Texas market is one of the most competitive in the United States, with dozens of options for consumers to choose from, and Texans are most aware of their alternatives. It is also important to note that without Texas, the awareness figures for the South would be significantly lower.

3.2. Consumer Price Sensitivity for Renewable Energy, Trended

According to Figure 8, consumer price sensitivity for renewable energy has increased over the past five years, in conjunction with overall increased price sensitivity for other green products, as well as the overall consumer marketplace. The change between 2006 and 2010 reveals a 9% annual decline in the number of consumers willing to spend between $5 and $20 more on renewable energy for their home.

In comparison to other green products, price sensitivity is actually lower for renewable energy. In 2010, 16% of the population reports that they would pay 20% more for products made in an environmentally-friendly and sustainable way, whereas one-quarter will pay $5–$20 more each month for renewable energy, which covers the premium in many cases.

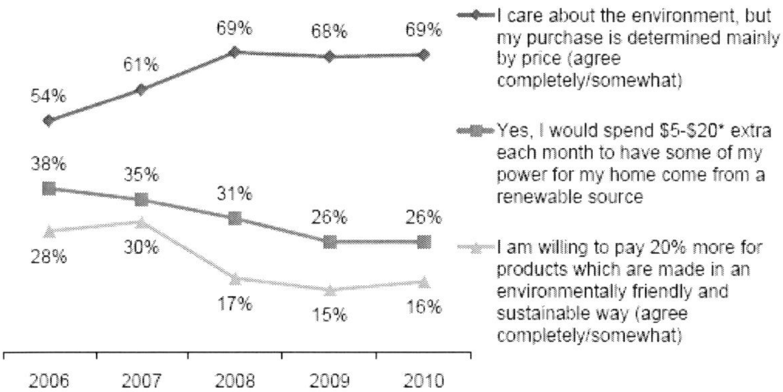

*$5–$10 prior to 2009.
Source: 2010 LOHAS Consumer Trends Database.

Figure 8. In Q.122, percent of the general population indicating their price sensitivity regarding renewable energy (trended); Q31, percent of the general population agreeing with each statement.

These figures actually far exceed the number of consumers who indicated that they currently buy renewable energy (7%).

While not everyone who says that they will pay more for renewable energy will follow through on those actions, it still suggests that there is room for market expansion.

3.3. Consumer Price Sensitivity for Renewable Energy, by Region

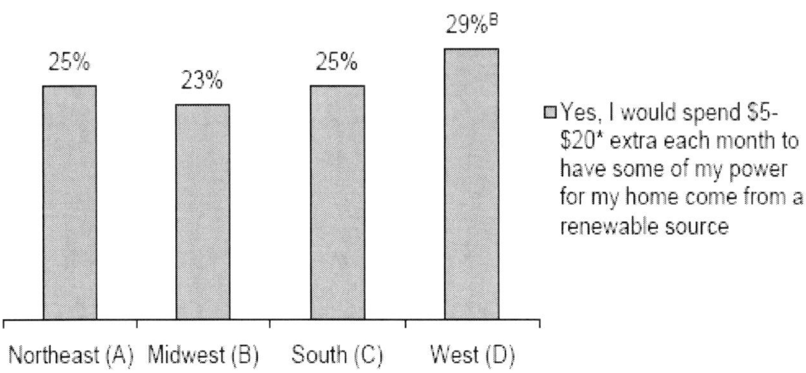

*$5–$10 prior to 2009.
Note: Capital letters indicate significant differences between regions at the 95% confidence level.
Source: 2010 LOHAS Consumer Trends Database.

Figure 9. In Q.122, percent of the general population, by region, stating that they would spend $5– $20 extra each month ($5–$10 prior to 2009) to have some of their household power come from a renewable source.

As shown in Figure 9, West Coast consumers are less price sensitive than those consumers in the eastern regions and statistically less price sensitive than those in the Midwest.

While these data may suggest that renewable energy marketers could charge more in the West than elsewhere in the country, it may also suggest that this region has higher market potential.

3.4. Consumer Stated Purchase of at Least Some of Their Household Power from Renewable Sources Over the Past Two Years

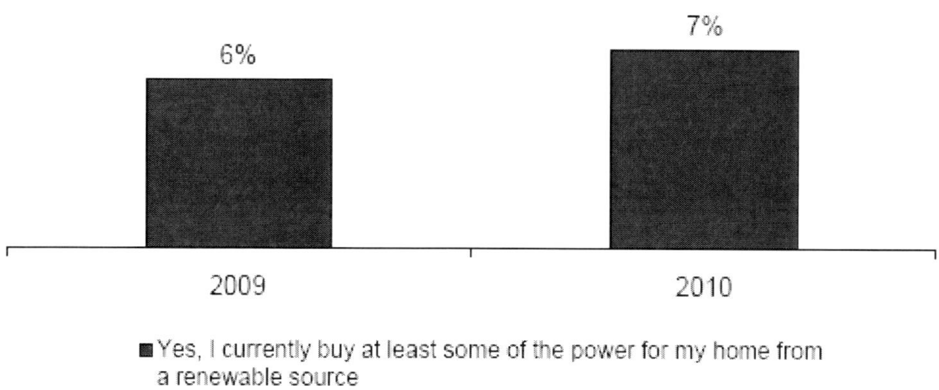

Source: 2010 LOHAS Consumer Trends Database.

Figure 10. In Q.122, percent of the general population stating that they currently buy at least some of their household power from a renewable source.

As shown in Figure 10, 7% of the population reports that they buy at least some of their household power from a renewable source. This is significantly higher than the percentage of consumers who participate in programs offered by consumer utilities and competitive marketers (approximately 1% of households). However, survey data may include individuals who know that renewable energy is part of the grid's mix in their service territory, who have solar-powered garden lights, solar water heating, solar electric systems, or other on-site renewable energy systems.

As other studies have found, this percentage is far below the portion of the population who care about renewable energy or who would pay more for it. Importantly, it is not that different from the percentage of people who know that they have the option to buy renewable energy. Simply providing consumers with greater awareness of their purchase options may be the most important factor in growing the renewable energy market.

Usage of other "green" market products may be helpful for comparative purposes (though the purchase process for these products is quite different than that for renewable energy). For example, one-quarter of the population has purchased a natural cleaning product, nearly 50% have purchased a CFL, and 30% have purchased rechargeable batteries in the past year.

As shown in Figure 11, Midwestern survey data indicates that consumers in this region are statistically less engaged than those in the West, with just 5% stating that they buy some of their power from a renewable source. This is not surprising, since the consumers in this region are the most price sensitive and the least aware of their purchase options.

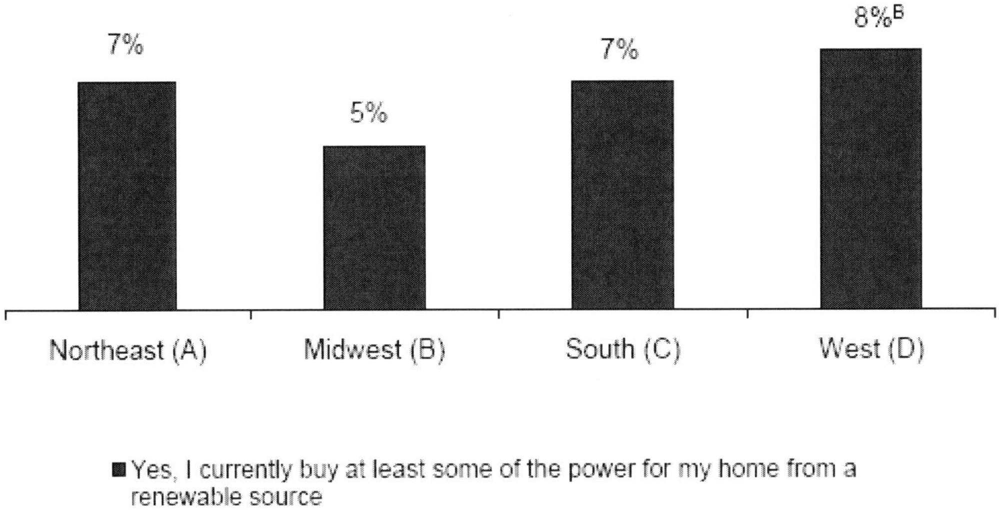

Capital letters indicate significant differences between regions at the 95% confidence level.
Source: 2010 LOHAS Consumer Trends Database.

Figure 11. In Q.122, percent of the general population, by region, stating that they currently buy at least some of their household power from a renewable source.

4. CONCLUSION

Overall, consumers have a high degree of awareness and concern for renewable energy, but few are aware of their purchase options or follow through on their stated concern, as indicated in Figure 12.

Major findings of this report include the following:

- Concern about renewable energy is quite high, with 80% of the general population caring completely or somewhat about the use of renewable energy sources. Although concern has decreased from 89% in 2002, such decline is consistent with a general decline in caring about environmental attributes, such as protecting the environment and sustainable agricultural practices.
- Consumers continue to believe that the most important benefit of renewable energy is that it is better for the environment than regular power, although some consumers believe that the most important benefits are domestic sourcing and to a lesser extent human health or economic benefits.

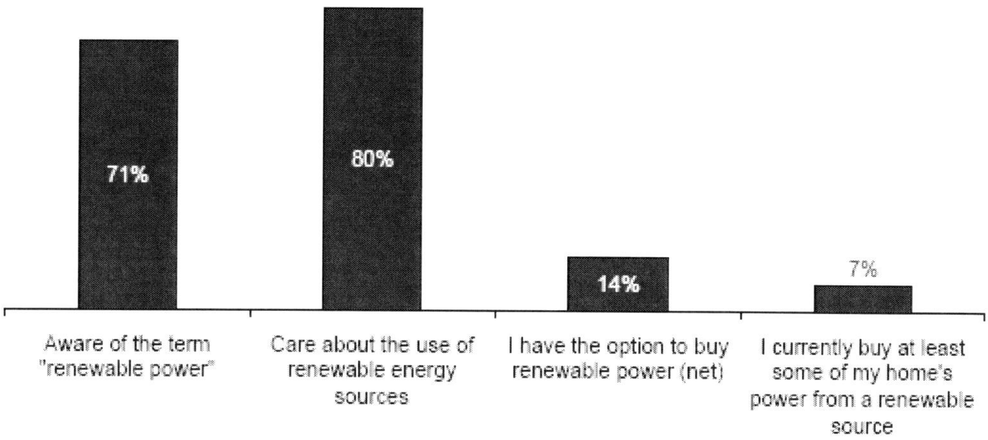

Source: 2010 LOHAS Consumer Trends Database.

Figure 12. Summary of consumer interest in renewable energy (2010 data).

- Consumer awareness is a challenge for utilities or companies providing renewable energy options to consumers. Despite a modest increase in consumer awareness, only 14% of consumers were aware of renewable energy purchase options in 2010. In addition, consumer willingness to pay more for renewable energy has declined in recent years.
- Despite the common perception that there are regional differences in attitudes about renewable energy, these data generally do not demonstrate strong regional differences, although the following differences do exist:
    - carbon footprint, renewable power, and carbon offset than consumers in other regions who have equivalent awareness. Consumers in the West are also less price sensitive than consumers in other regions and are more likely to report buying at least some of their household power from renewable sources.
    - Consumers in the Midwest are the least likely to be aware of their renewable power purchase options, while consumers in Texas and the Northeast, where retail competition exists and multiple alternative providers operate, are most aware of renewable energy purchase options.

Given low levels of product awareness, and that half of consumers who are aware of options to buy renewables do say they purchase renewables, these data suggest further opportunity for market growth in this sector, particularly if heightened consumer awareness can be achieved.

APPENDIX A. FULL SURVEY QUESTION TEXT

The questions reported in this paper appeared as follows in the 2010 LOHAS Consumer Trends Database. In some cases, other measures were included in the survey but not reported here, and those measures are not listed below.

9. Please indicate your level of agreement or disagreement with the following statements in terms of your own personal beliefs. *Please select one rating for each statement.* [*randomize*]

I care about....		Agree Completely	Agree Somewhat	Neither Agree Nor Disagree	Disagree Somewhat	Disagree Completely	Don't Understand/ Never Heard Of
a	Protecting the environment						
b	Sustainable agricultural practices						
c	Use of renewable energy sources						

14. Which of the following terms are you aware of? *Please select all that apply.* [*randomize*]

1. Biodegradable
2. Carbon footprint
3. Carbon offset
4. Sustainability

31. Thinking about *how* products are made and sold, please indicate your level of agreement or disagreement with the following statements in terms of how they affect your *purchasing* of products. *Please select one rating for each statement.* [*randomize*]

	When it comes to purchasing products or services...	Agree Completely	Agree Somewhat	Neither Agree Nor Disagree	Disagree Somewhat	Disagree Completely
a	I am willing to pay 20% more for products which are made in an environmentally friendly and sustainable way.					
b	I care about the environment, but my purchase is determined mainly by price.					

63. Which of the following *types* of products have you or someone in your household purchased in the past 12 months? *Please select all that apply.* 1. Compact fluorescent light bulbs (CFLs)

117. Please indicate the frequency with which you do the following: *Please check one for each.* [*Randomize*]

		Daily	Weekly	Monthly	Less Than Monthly	Never
a.	Not buy products from a brand or company that has practices I don't like					
b.	Take my own bag to the grocery store or other stores I shop frequently					

122. For the purposes of this survey, "renewable power" means electricity supplied from renewable energy sources, like wind and solar power. These energy sources are considered renewable sources because they are continuously replenished on the Earth. Renewable power can be used in exactly the same way as regular power. *Please select one response for each of the following.* [*randomize*]

		Yes	No	Don't Know
a	I have the option to buy renewable power from my electric/utility company.			
b	I have the option to buy renewable power from someone other than my current electric/utility company.			
c	I would spend $5–$20 extra each month to have some of the power for my home come from a renewable source.			
d	I currently buy at least some of the power for my home from a renewable source.			

123. What do you think is the most important benefit of renewable power? *Please select one.* [*randomize*]

1. Better for the environment than regular power
2. A source of energy from the United States
3. Better for human health than regular power
4. Better for the economy than regular power
5. None of the above [anchor]

122. For the purposes of this survey, "renewable power" means electricity supplied from renewable energy sources, like wind and solar power. These energy sources are considered renewable sources because they are continuously replenished on the Earth. Renewable power can be used in exactly the same way as regular power. Please select one response for each of the following. [randomize]

ACKNOWLEDGMENTS

This paper was authored by Gwynne Rogers at Natural Marketing Institute, an international strategic consulting, market research, and business development company specializing in the health, wellness, and sustainability marketplace.

This work was funded by the US Department of Energy's (DOE's) Office of Energy Efficiency and Renewable Energy (EERE). Lori Bird and Jenny Sumner of NREL served as technical monitors. The author wishes to thank Linda Silverman and the EERE technology programs for their support of this work.

The author also wishes to thank Blaine Collison of the US Environmental Protection Agency; Heidi Eggert of the Bonneville Environmental Foundation; Bridgett Neely of Green Mountain Energy; Orrin Cook, Jennifer Martin, and Alex Pennock of the Center for Resource Solutions; and Lynn Billman, Gail Mosey, and Robin Newmark of NREL for their thoughtful review of the document, as well as Sheri Anstedt for her editorial support.

REFERENCES

Bird, L.; Sumner, J. (September 2010). *Green Power Marketing in the United State: A Status Report (2009 Data).* NREL/TP-6A20-49403. Golden, CO: National Renewable Energy Laboratory, 69 pp.

Farhar, B. (July 1999). *Willingness to Pay for Renewable Electricity: A Review of Utility Market Research. Topical Issues Brief.* NREL/TP-550-26148. Golden, CO: National Renewable Energy Laboratory, 28 pp.

Lehr, R.L.; Guild, W.; Thomas, D.L.; Swezey, B.G. (June 2003). *Listening to Customers: How Deliberative Polling Helped Build 1,000 MW of New Renewable Energy Projects in Texas.* NREL/TP-620-33177. Golden, CO: National Renewable Energy Laboratory, 27 pp.

Natural Marketing Institute. (2002-2010). *Lifestyles of Health and Sustainability (LOHAS) Consumer Trends Database*. Harleysville, PA.

End Notes

[1] The Opinion Research Corporation 2007 poll found that 55% of consumers were either very or somewhat familiar with the term renewable energy. The poll also found that 16% of consumers were aware of a renewable energy purchase option through their utility company or electricity supplier. Of those consumers who were aware of an option to purchase, 29% reported that they did in fact purchase renewable energy through their utility company or electricity supplier.

[2] Compound annual growth rate (CAGR) is reported throughout this document. The CAGR reports growth from the first year to the most current year as if annual rates had changed at a steady rate and thus reduces the effects of year-to-year fluctuations.

[3] Renewable energy is a little broader than power, and there are forms of renewable energy that do not generate electricity, such as solar hot water.

[4] The quintessential high involvement category is a car or a house, where the purchase decision is very involved and impacts daily life. Consumer packaged goods are also a much more involved category because consumers have a more personal interaction with them (e.g., ingesting food or putting the product on the body), and the purchase cycle is rapid, meaning they are making these decisions on a weekly basis or even more often.

In: The Consumer Component in Renewable Energy Use ISBN 978-1-62081-425-3
Editors: Georgia M. Lee and Russ L. Cooper

Chapter 2

GREEN POWER MARKETING IN THE UNITED STATES: A STATUS REPORT (2009 DATA)*

Lori Bird and Jenny Sumner

LIST OF ACRONYMS

ACP	alternative compliance payment
aMW	average megawatt
CARB	California Air Resources Board
CCAR	California Climate Action Registry
CCX	Chicago Climate Exchange
DOE	Department of Energy
eGRID	Emissions & Generation Resource Integrated Database
EIAEnergy	Information Administration
EPA	Environmental Protection Agency
ERCOT	Electric Reliability Council of Texas
FEMP	Federal Energy Management Program
FPL	Florida Power & Light
GHG	greenhouse gas
GRP	General Reporting Protocol
kWh	kilowatt-hour
MW	megawatt
MWh	megawatt-hour
NREL	National Renewable Energy Laboratory
PJM-GATS	PJM-Generation Attribute Tracking System
REC	renewable energy certificate
RES	renewable energy standard
RGGI	Regional Greenhouse Gas Initiative

* This is an edited, reformatted and augmented version of a National Renewable Energy Laboratory Technical Report NREL/TP-6A20-49403 publication, Contract No. DE-AC36-08GO28308, dated September 2010.

RPS	renewable portfolio standard
SREC	solar renewable energy certificate
WECC	Western Electric Coordinating Council

EXECUTIVE SUMMARY

This report documents green power marketing activities and trends in the United States. Aggregate green power sales data for all voluntary purchase markets across the United States are presented for 2009. The data presented in this report are based primarily on figures provided to NREL by utilities and independent renewable energy marketeters. Because data cannot be obtained from all market participants, the estimates presented here likely represent an underestimate of the market size. Key trends identified in this year's report include:

- In 2009, total retail sales of renewable energy in voluntary markets exceeded 30 million MWh, an increase of 17% from 2008. The increase was dominated by renewable energy certificate (REC) sales, primarily to nonresidential consumers, which increased by about 20% from 2008 (see Figure ES-1). REC markets now represent 62% of total voluntary green power market sales.
- Utility green pricing programs in regulated electricity markets continued to grow on a sales basis but at a slower rate than in previous years, with sales volume increasing by about 7% in 2009. A relatively small number of utility programs continued to dominate sales and customer numbers. Utility premiums for green pricing continued to fall due in part to the increased cost competitiveness of renewable with conventional generation.
- Wind energy provided 73.7% of total green power sales volume, followed by biomass energy sources including landfill gas (10.0%), hydropower (9.9%), geothermal (0.2%), and solar (0.1%), with the remainder unknown (5.9%).
- Overall, the total number of customers purchasing green power increased by 44% in 2009, which is a higher rate than in previous years and with gains coming primarily from a competitive offering in Texas introduced in 2009. Utility green pricing program participants remained essentially flat in aggregate from 2007 to 2009, with some programs continuing to report customer losses in 2009.

In 2009, nearly 340,000 metric tons of avoided CO_2e from renewable energy facilities were marketed as offsets, an increase of approximately 39% from 2008. This is the equivalent of about 485,000 MWh of renewable energy generation.

INTRODUCTION

Voluntary consumer decisions to buy electricity supplied from renewable energy sources represent a powerful market support mechanism for renewable energy development. In the early 1990s, a small number of US utilities began offering "green power" options to their customers.[1] Since then, these products have become more prevalent, offered by traditional utilities and renewable energy marketers operating in states that have introduced competition into their retail electricity markets or offering renewable energy certificates (RECs) online. Today, more than half of all US electricity customers have an option to purchase some type of green power product directly from a retail electricity provider, while all consumers have the option to purchase RECs.

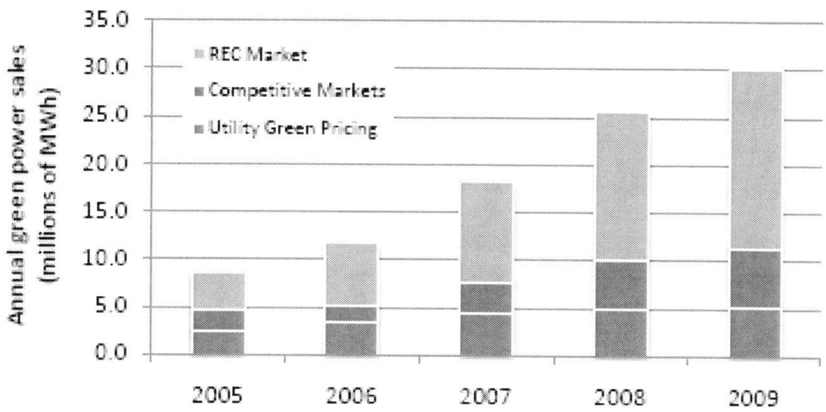

Figure ES-1. Estimated annual green power sales by market sector, 2005–2009.

Approximately 860 utilities, or more than 25% of utilities nationally, offer green power programs to customers. These programs allow customers to purchase some portion of their power supply as renewable energy—almost always at a higher price—or to contribute funds for the utility to invest in renewable energy development, such as solar installations at local schools. The term "green pricing" is typically used to refer to these utility programs offered in regulated or noncompetitive electricity markets.

In states with competitive (or restructured) retail electricity markets, electricity customers can often buy electricity generated from renewable sources by switching to an alternative electricity supplier that offers green power. In some of these states, default utility electricity suppliers offer green power options to their customers in conjunction with competitive green power marketers.[2]

A dozen states that have opened their markets to retail competition have experienced some green power marketing activity.[3]

Finally, regardless of whether they have access to a green power product from their retail power provider, energy consumers can purchase green power through RECs, which represent the environmental attributes of electricity generated from renewable energy-based projects. Consumers can also support renewable energy development through REC purchases without having to switch to an alternative electricity supplier. Today, several dozen companies actively market RECs to residential or business customers throughout the United States. Some REC marketers also sell greenhouse gas (GHG) emissions offsets sourced from renewable energy projects.

This report documents green power marketing activities and trends in the United States. First, aggregate green power sales data for all voluntary purchase markets across the United States is presented. The next three sections provide summary data on 1) utility green pricing programs offered in regulated electricity markets; 2) green power marketing activity in competitive electricity markets, as well as green power sold to voluntary purchasers in the form of RECs; and 3) renewable energy sold as GHG offsets in the United States. These sections are followed by a discussion of key market trends and issues. The final section offers conclusions and observations.

The data presented in this report are based primarily on figures provided to NREL by utilities and independent renewable energy marketers.[4] NREL also supplements this data with information from REC certifiers, REC tracking systems (see ERCOT 2009), and press releases describing large voluntary green power purchases. Because data cannot be obtained from all market participants, the estimates presented here likely represent an underestimate of the market size.

Data on the competitive markets is particularly challenging to obtain due to market sensitivity and rapid changes in offerings, and therefore estimates of the competitive market are more uncertain.

GREEN POWER MARKET SUMMARY AND TRENDS

Green Power Sales

Green power sales, driven by REC markets, increased by 17% to approximately 30 million MWh from 2008–2009.

Overall, retail sales of renewable energy in voluntary purchase markets exceeded 30 million MWh in 2009, or about 0.8% of total US electricity sales.[5] Estimates presented in this report are primarily based on data provided by utilities and marketers and supplemented with other available data.[6] Because we are unable to obtain data from all market participants, the estimates presented here likely underestimate the size of the entire market.

In addition to renewable energy sales, GHG offsets sourced from "new" renewable energy resources—totaling 339,200 metric tons of CO_2 equivalent (CO_2e)—were sold to US voluntary purchasers in 2009. Generation from a renewable energy source can generally be sold either as a MWh or as a metric ton of CO_2e.

Wind energy represented 73.7% of total green power sales, followed by biomass energy sources including landfill gas (10.0%), hydropower (primarily low impact or small hydro; 9.9%), geothermal (0.2%), solar (0.1%), and unknown sources (5.9%) (see Figure 1). Based on the sales data presented in this report, we estimate the market value of green power sales (the above-market cost of the green power) in 2009 to be between $136 million and $236 million.[7]

Green power sales (in megawatt-hours) increased by 17% in 2009 from 2008, with a compound annual growth rate of 37% since 2005 (see Table 1 and Figure 2). REC sales have been driving much of the growth, increasing 20% in 2009 from 2008. Overall, REC markets represent 62% of all green power sales.[8] Annual growth rates in all market sectors declined in 2009 compared to 2008.

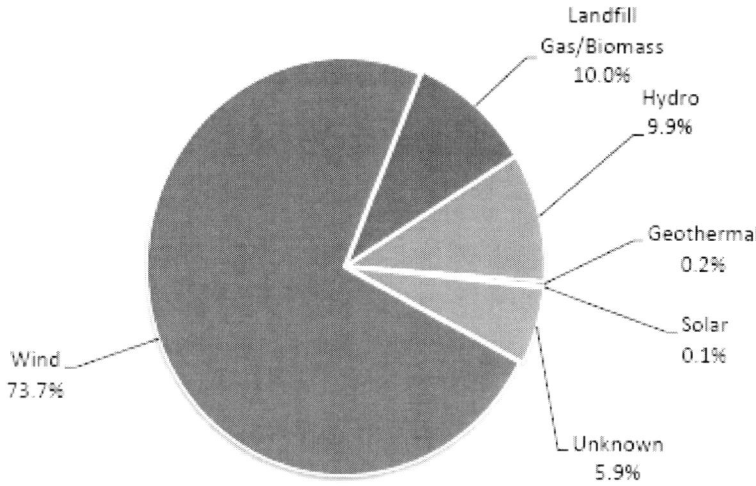

Figure 1. Estimated green power sales by renewable energy source, 2009.

Sales to nonresidential customers continued to outpace those to residential customers, with more than 76% of all sales by volume to the nonresidential sector in 2009, an increase from 65% in 2005 (see Table 2). Nearly all REC sales were to business and institutional customers, while residential customers played a larger role in green pricing programs and competitive markets, where they accounted for 54% and 69%, respectively, of renewable energy sales (see Table 3).

Table 1. Estimated Annual Green Power Sales by Market Sector, 2005–2009[a] (Millions of MWh)

Market Sector	2005	2006	2007	2008	2009	% Change	% Change	% Change	% Change
						'05–'06	'06–'07	'07–'08	'08–'09
Utility Green Pricing	2.5	3.4	4.2	4.8	5.2	39%	23%	15%	7%
Competitive Markets	2.2	1.7[b]	3.2	5.3[d]	6.2	-20%[b]	88%[b]	66%[d]	18%
REC Markets[c]	3.9	6.8	10.6	15.6	18.7	75%	55%	47%	20%
Retail Total	*8.5*	*11.9*	*18.0*	*25.7[d]*	*30.0*	*41%*	*51%*	*43%[d]*	*17%*

[a] Includes sales of new and existing renewable energy. Totals and growth rates may not compute due to rounding.

[b] 2006 sales figures may be underestimated because of data gaps.

[c] Includes only RECs sold to end-use customers separate from electricity.

[d] 2008 competitive market sales were revised upward in this report to reflect data on green power markets in Texas published by the Texas Public Utilities Commission in 2010.

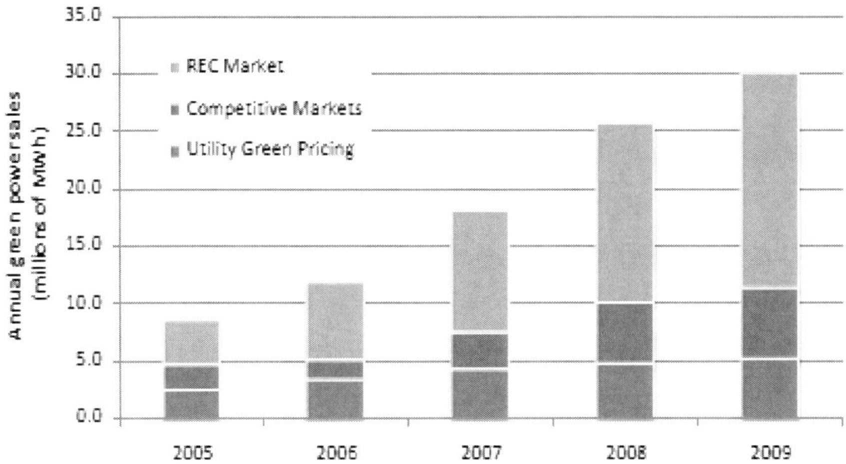

Figure 2. Estimated annual green power sales by market sector, 2005–2009.

Table 2. Estimated Annual Green Power Sales by Customer Segment, 2005–2009 (Millions of MWh)

Customer Segment	2005	2006	2007	2008	2009	% Change '05–'06	% Change '06–'07	% Change '07–'08	% Change '08–'09
Residential	3.0	3.2	4.5	6.5	7.2	8%	39%	43%	12%
Nonresidential	5.5	8.7	13.6	19.2	22.8	58%	56%	41%	19%
Total	*8.5*	*11.9*	*18.1*	*25.7*	*30.0*	*41%*	*53%*	*42%*	*17%*
% Nonresidential	65%	73%	75%	75%	76%	--	--	--	--

Note: Totals and growth rates may not add or calculate due to rounding.

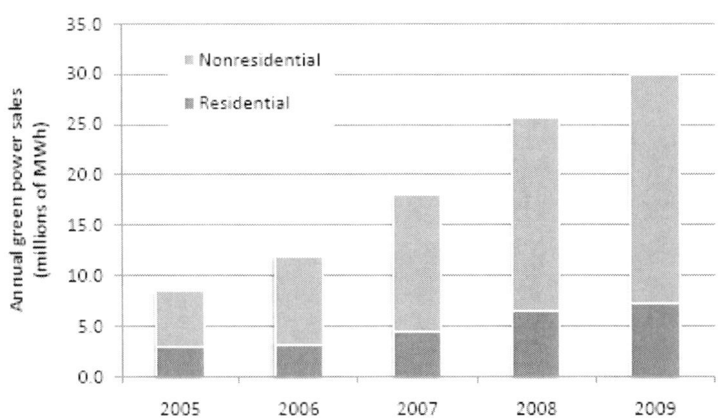

Figure 3. Residential and nonresidential green power sales, 2005–2009.

Table 3. Estimated Annual Green Power Sales by Customer Segment and Market Sector, 2009 (Millions of MWh)

Customer Segment	Green Pricing	Competitive Markets	REC Markets	Total
Residential	2.8	4.3	0.04	7.2
Nonresidential	2.3	1.9	18.6	22.8
Total	5.2	6.2	18.7	30.0
% Residential	54%	69%	0.2%	24%

Note: Totals may not add due to rounding.

Table 4. Estimated Cumulative Renewable Energy Capacity Supplying Green Power Markets, 2006–2009 (MW)

Market	2006 Total RE Capacity	2006 New RE Capacity	2007 Total RE Capacity	2007 New RE Capacity	2008 Total RE Capacity	2008 New RE Capacity	2009 Total RE Capacity	2009 New RE Capacity
Utility Green Pricing	1,100	1,000	1,400	1,300	1,500	1,400	1,700	1,600
Competitive Markets/RECs	2,400	2,100	3,700	3,000	5,800	4,900	7,700	6,400
Total	3,500	3,100	5,100	4,300	7,300	6,300	9,400	8,000

Note: "New" renewable energy capacity is a subset of total renewable energy capacity supplying green power markets.

At the end of 2009, megawatt-hour sales of renewable energy in voluntary markets represented a generating capacity equivalent of about 9,400 MW, with about 8,000 MW of that from new renewable energy sources (see Table 4).[9,10] Since 2006, the amount of renewable energy capacity serving green power markets increased nearly threefold.

Customer Participation

Participation in REC markets and utility green pricing programs remained relatively flat; one competitive program pushed competitive market participation up by 110%.

Based on the information we have obtained, we estimate that approximately 1.4 million electricity customers nationwide purchased green power products in 2009 through regulated utility companies, from green power marketers in a competitive-market setting, or in the form of RECs (see Table 5).[11] Participation in REC markets and utility green pricing programs remained relatively flat while competitive market participation increased about 110% primarily because of substantial customer increases reported by one competitive marketer.

Up until 2007, utility green pricing programs showed continued customer growth as the number of utility programs increased and as existing programs grew; however, in 2008, participation was essentially flat, largely due to the cancellation of the Florida Power and Light (FPL) Sunshine Energy Program, a large program with more than 35,000 participants prior to its termination. In 2009, customer numbers continued to remain flat—growth was 1% from 2008 to 2009.

Competitive market green power participation expanded considerably in 2009 as a result of increased participation reported by one marketer in Texas. While the number of green power purchasers has expanded during the past few years in markets with retail competition, participation has been less consistent over time, as some markets have grown and then contracted (such as in California and Pennsylvania). In the last few years, growth in competitive markets has been concentrated in Texas and a few programs in the Northeast.

In 2009, the number of customers buying RECs declined from 2008 but remained greater than in previous years (2003–2007).

The number of customers buying RECs still represents a small fraction of the total green power market on a customer basis but not on a kilowatt-hour basis.

Despite the limited number of residential customers purchasing RECs, REC sales represent 62% of green power kilowatt-hour sales (see Table 3) and have grown dramatically in recent years as a result of large purchases by nonresidential customers (see Appendix B for a list of top green power purchasers). Average participation rates among utility green pricing programs decreased slightly from 2.2% in 2008 to 2.0% in 2009, with a median value of 1.0%; top performing programs achieved rates ranging from 5.1% to 20.8%. Competitive markets experienced green power customer penetration rates ranging from 1.7% to 2.5% in the states with the most active markets, and in Texas, participation in competitive markets at the state level is much higher at more than 4.5%. Participation in competitive markets has been subject to market conditions and rules and has been more volatile than in traditionally regulated markets.

Table 5. Estimated Cumulative Green Power Customers by Market Segment, 2003–2009

	2003	2004	2005	2006	2007	2008	2009
Utility Green Pricing Customers	270,000	330,000	390,000	490,000	550,000	550,000	550,000
Competitive Market Customers	>170,000	>140,000	>180,000	~210,000	300,000	390,000	830,000
REC Market Customers*	< 10,000	< 10,000	< 10,000	~10,000	>10,000	30,000	< 20,000
Retail Total	*-450,000*	*-480,000*	*-580,000*	*-710,000*	*-860,000*	*-970,000*	*-1,400,000*
% Change	*-15%*	*-7%*	*-21%*	*-22%*	*-21%*	*-13%*	*-44%*

*Includes only end-use customers purchasing RECs separate from electricity.

Note: In some cases, estimates have been revised from those reported in previous NREL reports as updated data have become available.

Note: Totals may not add due to rounding.

Comparison of Voluntary and Compliance Markets

Compliance demand for new renewable energy was approximately equivalent to voluntary demand.

In 29 states and Washington, D.C., renewable portfolio standard (RPS) policies require that utilities or load-serving entities include a certain percentage of renewable energy within their power generation mix; the percentages required and eligibility requirements vary among the states. Voluntary purchases of renewable energy are almost always in addition to renewable energy used to meet RPS targets.[12] Green power certification programs and state RPS policies generally ensure that there is no double counting between the two markets (i.e., that the same kilowatt-hour is not used for more than one purpose).[13] Ensuring the absence of double counting is important to the integrity of the market because consumers who pay a premium for green power want to support renewable energy that would not have been otherwise supported through regulatory requirements.

In 2009, state RPS policies collectively called for utilities to procure about 29.5 million MWh of new renewable energy generation (Barbose 2010) compared to about 30.0 million MWh sold into the voluntary green power market.[14] Figure 4 shows that between 2004 and 2008, voluntary market demand for renewable energy slightly exceeded compliance market demand for new renewable energy, while in 2009, compliance demand for new renewable energy was slightly greater than voluntary market demand. Renewable energy demand required to meet RPS policies is expected to grow rapidly in coming years. By 2010, RPS policies collectively call for utilities to obtain approximately 52 million MWh of new renewable energy, increasing to more than 100 million MWh in 2014; voluntary market growth rates would have to increase to keep pace.

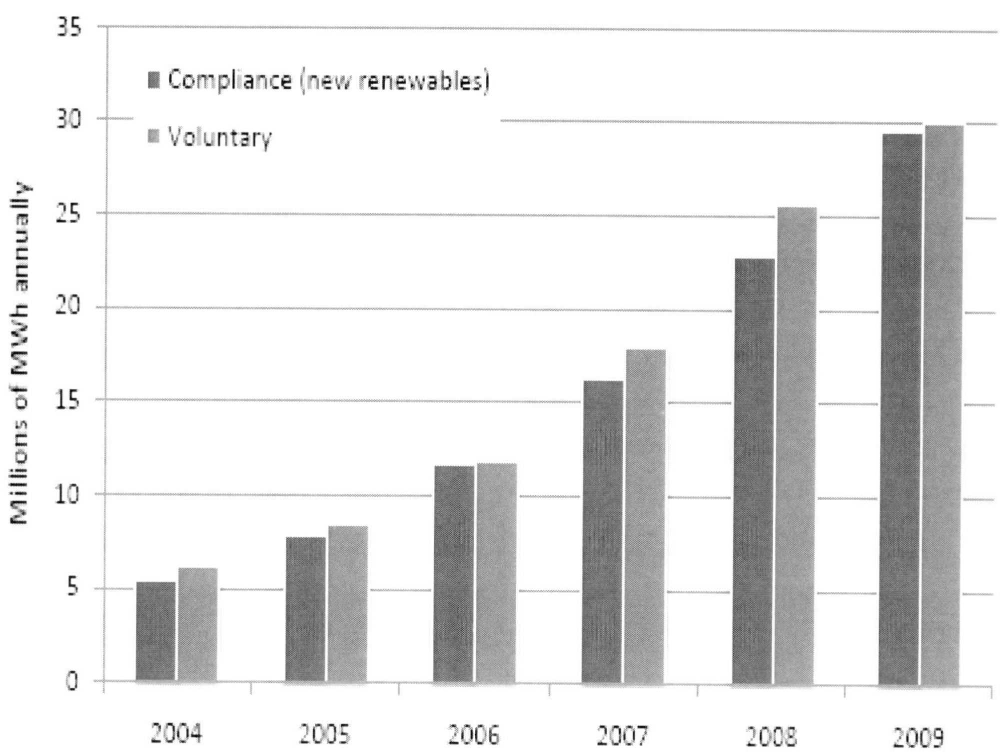

Note: Compliance market data sourced from Lawrence Berkeley National Laboratory (Barbose 2010).

Figure 4. Comparison of compliance and voluntary markets for renewable energy, 2004–2009.

UTILITY GREEN PRICING

This section provides information specific to utility green pricing programs, a subset of the market. The number of utilities offering green pricing has grown steadily in recent years—in 2009, approximately 860 investor-owned, public, and cooperative utilities in most states offered green pricing programs.

Appendix D provides links to Web pages with a compilation of all green power product offerings, and Appendix E provides a list of utilities offering green pricing. Because a number of small municipal or cooperative utilities offer programs developed by their power suppliers, the number of distinct green pricing programs is about 160. Some states have adopted laws requiring utilities to offer consumers green power options, which have driven the development of new programs.[15]

Green Pricing Products and Premiums

Average green pricing premium continued to decrease.

Typically, green pricing programs are structured so that customers can either purchase green power for a certain percentage of their electricity use (often called "percent-of-use products") or in discrete amounts or blocks at a fixed price ("block products"), such as a 100 kWh block. Most utilities offer block products but may also allow customers to buy green power for their entire monthly electricity use. Utilities that offer percent-of-use products generally allow residential customers to elect to purchase 25%, 50%, or 100% of their electricity use as renewable energy, while a few offer fractions as small as 10%. Under these types of programs, larger purchasers, such as businesses, can often purchase green power for some fraction of their electricity use as well.

In 2009, the price of green power for residential customers in utility programs ranged from -0.17¢/kWh (a savings compared to standard service) to 10.00¢/kWh above standard electricity rates, with an average premium of 1.75¢/kWh and a median premium of 1.50¢/kWh. These premiums have been adjusted to account for any fuel-cost exemptions granted to green power program participants.[16]

In 2009, the 10 utility programs with the lowest premiums for energy derived from new renewable sources had premiums ranging from -0.17¢/kWh (a savings) to 0.80¢/kWh. On average, consumers spend about $5.40 per month above standard electricity rates for green power through utility programs, which is consistent with previous years.

Since 2000, the average price premium has dropped at a compound annual rate of 7% (Figure 3). Some of this reduction can be attributed to lower market costs for renewable energy supplies or increased competitiveness with conventional generation sources.

The competitiveness of wind and other renewables with conventional generation, as well as regional demand from state renewable energy standards (and national demand if a federal standard is adopted), will affect premiums in coming years.

Table 6. Residential Price Premiums of Utility Green Power Products, 2002–2009 (¢/kWh)

	2002	2003	2004	2005	2006	2007*	2008*	2009*
Average Premium	2.82	2.62	2.45	2.36	2.12	1.85	1.80	1.75
Median Premium	2.50	2.00	2.00	2.00	1.78	1.50	1.50	1.50
Range of Premiums	0.70-17.60	0.60-17.60	0.33-17.60	(0.70)-17.60	(0.10)-17.60	0.09-7.50	(1.00)-8.80	(0.17)-10.00
10 Programs with Lowest Premiums**	0.70-1.50	0.60-1.30	0.33-1.00	(0.70)-0.90	(0.10)-1.00	0.09-0.80	(1.00)-0.90	(0.17)-0.80
Number of Programs Represented	80	91	101	104	97	71	86	92

*In later years, calculations of premiums were based on programs that responded to the questionnaire. In previous years, a larger sample of programs was used to calculate the premium, as data were available.

**Represents the 10 utility programs with the lowest price premiums for new, customer-driven renewable energy. This includes only programs that have installed—or announced firm plans to install or purchase power from—new renewable energy sources.

GREEN PRICING CUSTOMER PARTICIPATION

Aggregate green pricing participation was largely flat; average program participation rates decreased slightly.

At the end of 2009, about 552,200 customers were participating in utility green pricing programs in regulated electricity markets (see Table 7).[17]

As in the past, a relatively small number of green power programs account for the majority of customers, with just 10 programs accounting for 75% of all participants (see Appendix E).[18] From 2001 to 2007, the number of customer participants increased more than threefold, but this trend reversed in 2008. With the cancellation of the large FPL program, nearly 40,000 customers left the market,[19] and total participants in utility programs nationwide fell slightly.

Without the loss of the FPL program, the number of participants in utility green power programs would have grown modestly in 2008, by about 6%. In 2009, customer participation remained relatively unchanged from 2008, growing just 1% (see Table 7).

The decline in the economy in the second half of 2008 and in 2009 likely contributed to smaller gains in participants relative to previous years, and a number of programs reported losses in the total number of participants. In 2009, residential participation increased by 1%, while nonresidential participation declined by 1%, whereas in previous years, total participation increased at a greater rate (see Table 7).

Of the 62 utility programs that reported participation data in both 2008 and 2009, 32 utilities (52%) saw net declines in participation, 28 utilities (45%) saw net gains in participation, and 2 utilities (3%) had the same number of participants.

Table 7 delineates residential and nonresidential customer participation in utility green pricing programs over time. The vast majority of participants are residential customers, with nonresidential customers accounting for only 4.7% of all participants.

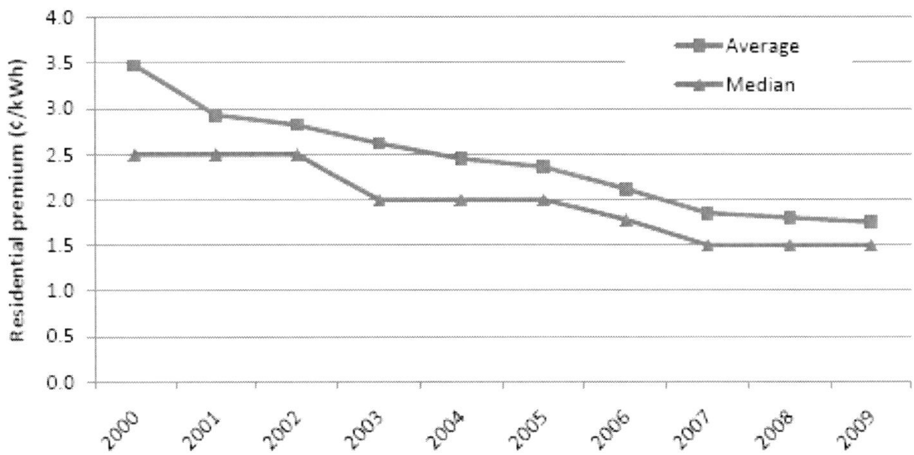

Figure 5. Trends in utility green pricing premiums, 2000–2009.

From 2002 to 2008, nonresidential participation was growing at a faster rate than residential participation; however, in 2009, this trend reversed, with nonresidential customers declining by 1% and residential customers increasing by 1%. At the end of 2009, the average participation rate in utility green pricing programs among eligible utility customers was 2% with a median of 1% (Table 8). These industry-wide rates have shown little change in recent years, though 2009 did see a decrease in participation rates, likely a result of the economic recession. Top-performing programs have demonstrated improvement, with participation rates ranging from about 5% to 21% in 2009, compared to a range of 3% to 6% in 2002, though participation rates in top performing programs have remained relatively unchanged since 2007. The 20% participation threshold was exceeded for the first time in 2007.

Table 7. Estimated Cumulative Number of Customers Participating in Utility Green Pricing Programs (Regulated Electricity Markets Only), 2002–2009

Customer Segment	2002	2003	2004	2005	2006	2007	2008	2009
Residential	224,500	258,700	323,700	383,400	470,800	526,700	519,700	526,300
Nonresidential	3,900	6,500	8,100	11,300	15,500	20,200	26,100	26,000
Total	*228,400*	*265,200*	*331,800*	*394,700*	*486,300*	*546,900*	*545,800*	*552,200*
% Total Annual Growth	35%	16%	25%	19%	23%	12%	0%	1%
% Residential Growth	35%	15%	25%	18%	23%	12%	-1%	1%
%Nonresidential Growth	56%	67%	25%	40%	37%	30%	29%	-1%

Note: Totals may not add due to rounding.

Table 8. Customer Participation Rates in Utility Green Pricing Programs, 2002–2009

Participation Rate	2002	2003	2004	2005	2006	2007	2008	2009
Average	1.2%	1.2%	1.3%	1.5%	1.8%	2.0%	2.2%	2.0%
Median	0.8%	0.9%	1.0%	1.0%	1.0%	1.3%	1.2%	1.0%
Top 10 Programs	3.0%–5.8%	3.9%–11.1%	3.8%–14.5%	4.6%–13.6%	5.1%–16.9%	5.2%–20.4%	5.0%–21.0%	5.1%–20.8%

In 2009, utilities reported that an average of 7.8% and a median of 6.3% of customers dropped out of green pricing programs, an increase from 2008 when utilities reported that an average of 5.5% and a median of 2.5% of customers dropped out. Although the average and median dropout rates are higher than in previous years, likely due to the economic recession, retention rates are still relatively high. This finding suggests that customers tend to be "sticky" and maintain participation in green power programs. While data on the reason for dropouts is not available, anecdotal evidence from some utilities suggests that customer moves can be a significant source of dropouts. Most utilities (about 66%) do not impose minimum periods for which customers must subscribe to the green power program. If a minimum term is imposed, it is most commonly one year—although there are several programs that offer fixed-price green power for contracts of longer durations, particularly to nonresidential customers.

Green Pricing Renewable Energy Sales

Green pricing sales increase modestly in 2009; average purchase size increased.

Utility green pricing sales continue to exhibit some growth, but growth has slowed in the past three years. Collectively, utilities in regulated electricity markets sold about 5.2 billion kWh of green power to customers in 2009 (Table 9). Green pricing program sales to all customer classes grew by 7% in 2009, compared to rates ranging from 15% to 43% in recent years (Table 9 and Figure 4).

Table 9. Annual Sales of Renewable Energy through Utility Green Pricing Programs (Regulated Electricity Markets Only), 2003–2009 (Thousands of kWh)

	2003	2004	2005	2006	2007	2008	2009
Residential	870	1,300	1,610	2,100	2,550	2,660	2,820
Nonresidential	410	540	840	1,300	1,630	2,150	2,320
Total Sales	*1,280*	*1,840*	*2,450*	*3,400*	*4,190*	*4,810*	*5,150*
% Annual Growth in Total Sales	43%	43%	33%	39%	23%	15%	7%
% Nonresidential of Total Sales	32%	30%	34%	38%	39%	45%	45%

Note: Totals may not add due to rounding.

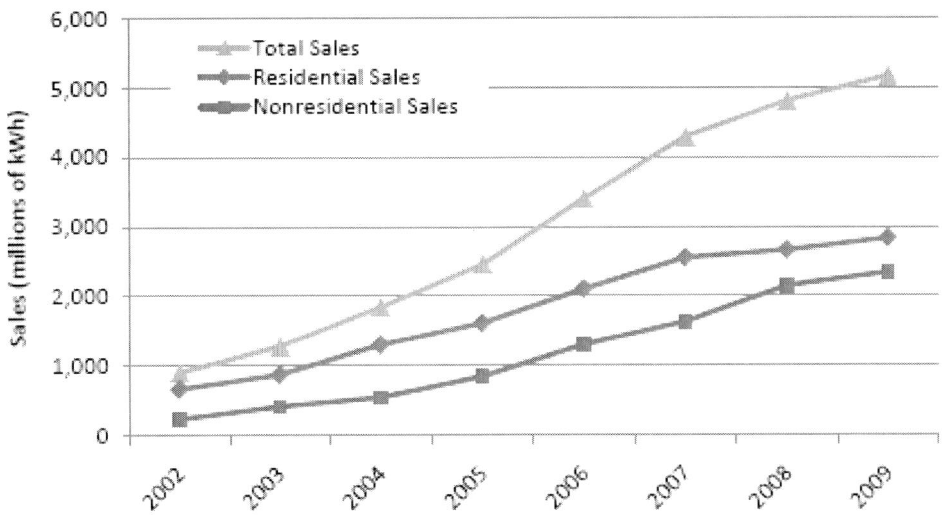

Figure 6. Annual sales of renewable energy through utility green pricing programs, 2002–2009 (regulated electricity markets only).

In 2009, the average residential purchase decreased slightly, while the average nonresidential purchase increased slightly after nearly doubling from 2007 to 2008 (Table 10). Although the reason for increased purchases by nonresidential customers is not known, it could be attributed to a decline in green power prices for nonresidential retail customers or enrollment of larger commercial and industrial customers. Some programs may have also placed greater emphasis on marketing to the commercial sector to make up for residential customer losses.

Table 10. Average Purchases of Renewable Energy per Customer, 2002–2009 (kWh/year)

	2002	2003	2004	2005	2006	2007	2008	2009
Residential	2,900	3,400	4,000	4,200	4,400	4,900	5,500	5,100
Nonresidential	60,000	63,100	67,200	74,500	85,700	77,400	141,300	146,300
All Customers	3,900	4,800	5,500	6,200	6,700	7,400	20,800	26,300

The vast majority (about 96%) of the renewable energy sold to consumers through green pricing programs was supplied from projects meeting the generally accepted industry definition of "new." Renewable energy sold through green pricing programs in 2009 represents an equivalent renewable energy capacity of 1,645 MW, with 1,585 MW of this represented by new renewable energy resources (Table 11).[20]

In 2008, sales of renewable energy through green pricing programs represented more than 1,500 MW of renewable energy capacity, with about 1,400 MW of that from new renewable energy sources.

Wind, solar, landfill gas, and other biomass are the renewable resources most commonly included in utility programs; although solar in particular may be used to supply a small fraction of MWh sales.

Wind energy represents the largest portion of the total capacity. Table 4 presents estimates of new capacity-serving green pricing programs in earlier years. In 2009, green power sales represented a small proportion of a utility company's overall energy sales. Table 12 shows that, on average, renewable energy sold through green pricing programs in 2009 represented 1.0% of total utility electricity sales (on a MWh basis). The average percentage of green power sold compared to total utility electricity sales in 2009 remained unchanged from 2008, while the median percentage dropped slightly. A few utilities reported fractions as high as 8% of total retail electricity sales, and, due to a large nonresidential purchase, one small utility reported 21.4% of total retail electricity sales (see Table E-4 in Appendix E). On a residential basis, green power sales represented a higher fraction of total utility electricity sales, with one utility reporting a fraction as high as 24.9%.

Table 11. Renewable Energy Generation and Capacity Supplying Green Pricing Programs, 2009

	Landfill Gas	Other Biomass	Geo-thermal	Hydro	Solar	Wind	Unknown	Total
Sales (MWh)	353,400	248,600	45,000	63,100	18,875	4,434,400	1,700	5,165,000
% of Total Sales	7%	5%	1%	1%	0.4%	86%	0.03%	100%
Total Sales (MW)	45	35	5	18	14	1,534	1	1,652
Total New Sales (MW)	42	35	5	17	13	1,472	-	1,585

Note: Totals may not add due to rounding.

Table 12. Renewable Energy Sales as a Percent of Utility Electricity Sales, 2008–2009

Customer Class		2008			2009	
	Avg.	Med.	Range	Avg.	Med.	Range
Residential	1.5%	0.5%	0%–23.4%	1.4%	0.4%	0%–24.9%
Nonresidential	0.8%	0.2%	0%–12.0%	1.2%	0.2%	0%–21.6%
All Customers	1.0%	0.4%	0%– 6.4%	1.0%	0.3%	0%–21.4%

Competitive Green Power and REC Markets

This section provides greater detail on green power sold in competitive (or restructured) retail electricity markets as well as in the form of RECs—subsets of the entire green power market. About one-quarter of US states have restructured their electricity markets for retail service competition. Currently, electricity consumers in the following states can purchase competitively marketed green power: Connecticut, Delaware, Illinois, Maine, Maryland, Massachusetts, Michigan, New Jersey, New York, Pennsylvania, Rhode Island, Texas, and D.C.21,22 Competitively marketed green power offerings are also available to nonresidential consumers in a few other states.

Initially, buying green power in competitive retail markets entailed switching electricity service from the incumbent utility to a green power supplier. In some markets, there was limited switching, and as a remedy, a number of states now require default suppliers (which are often the incumbent distribution utilities) to offer green power options to their customers. These load-serving entities typically provide customers with underlying electricity generation, combined with a choice of several green products offered by competing green power marketers. In addition, several utility suppliers have voluntarily teamed with a single green power marketer to offer a green power option to their customers. Such programs are now offered in Connecticut, Massachusetts, New Jersey, New York, Pennsylvania, and Rhode Island.

In addition to competitive offerings, RECs provide another alternative to switching electricity suppliers. Also known as green certificates, green tags, or tradable renewable certificates, RECs represent the "green" attributes of renewable energy generation and can be sold separately from commodity electricity. REC-based products may be supplied from a variety of renewable energy sources throughout the country and sold to customers nationally, or they may be supplied from renewable energy sources in a particular region or locality and marketed as such to local customers. More than 25 companies offer certificate-based green power products to retail customers via the Internet, and a number of other companies market RECs solely to commercial and industrial customers.[23]

RECs are also sold in the wholesale market and are frequently used by utilities and marketers who bundle RECs with commodity electricity to sell green power to retail customers. In fact, RECs are used to supply most of the programs where default suppliers have teamed with green power marketers. Therefore, it can be difficult to distinguish REC products from other green power offerings. This is particularly true when REC products are supplied from renewable sources located in the same region where they are marketed.

REC and Competitive Market Products and Pricing

On the whole, retail REC products continued to be less expensive than competitive market products; 73% of the total kilowatt-hours sold in the retail voluntary market in 2009 were involved in a Green-e Energy certified transaction at some point in their chain of custody.

Green power products offered in electricity markets with retail competition tend to differ from those offered by utilities in regulated markets, as they are more likely to be sourced from RECs because suppliers may be less able to enter into long-term contracts with generators. In addition, price premiums may fluctuate more frequently.

Initially, green power marketers in competitive markets were often forced to offer existing renewable energy sources because of a lack of new renewable energy supplies, but most marketers now offer primarily new renewable energy. In 2009, about 81% of competitive-market and REC sales were supplied from new renewable energy sources. This movement toward increased reliance on new sources has also been encouraged by green power product certification programs, which set standards for product quality and have required increasing amounts of new renewable energy. Beginning January 1, 2007, the Green-e Energy[24] certification program began requiring that all certified products be supplied exclusively from new renewable energy projects. The US Environmental Protection Agency's (EPA) Green Power Partnership requires its partners to purchase new renewable energy to meet its purchase criteria.[25] Currently, both the Green-e Energy National Standard and EPA define new as those facilities put into service on or after January 1, 1997, which is generally considered to be the inception of the voluntary green power market. Beginning on July 15, 2011, the Green-e Energy National Standard will have a 15-year rolling "new date," meaning that projects must have come online within 15 years prior to the sale of the green power in order to be classified as new.

The price premium charged for competitive-market products depends on several factors including the price of standard offer or default service, the availability of incentives to green power marketers or suppliers, and the cost of renewable energy generation available in the regional market. Some marketers have charged prices close to or even below the default market price in recent years (e.g., in Texas); others have offered fixed-price products, providing customers with protection against increasing prices for a specified period of time, usually one year.

Competitively marketed green power products generally carry a price premium between 1.3¢/kWh and 3.7¢/kWh for residential and small commercial customers, although offerings have ranged from small discounts to a premium of about 10¢/kWh in recent years. For utility/marketer programs offered in states with retail competition, the average price premium for green power was about 2.2¢/kWh. In addition, price premiums can change frequently with changes in market conditions. Higher-priced products often contain a larger fraction of new renewable energy content or resources that are more desirable to consumers, such as new wind and solar. Retail prices charged for REC products typically range from about 1¢/kWh to 2.5¢/kWh for residential and small commercial customers, although some are priced as high as 10–20¢/kWh for some products, such as solar RECs.

In most cases, large commercial customers are able to negotiate lower prices. Nearly all REC products are sourced from new renewable energy generation projects as a result of product certification requirements. REC buyers often seek certification out of concerns over double counting and to ensure a level of oversight and auditing because RECs are generally not subject to the same regulatory scrutiny as electricity and mandatory renewable requirements. Table 13 shows *Green-e Energy* certified retail transactions in 2008 and 2009. *Green-e Energy* certified more than 18.6 million MWh of retail transactions in 2009 (Terada 2010). Compared to NREL's total voluntary market retail sales figure of 30.0 million MWh, *Green-e Energy* certified 62% of voluntary market retail sales.

The *Green-e Energy* program also certifies wholesale renewable energy transactions, which exceeded 8.9 million MWh in 2009. It is important to note that 5.7 million MWh sold in certified wholesale transactions were resold in *Green-e Energy* certified retail transactions. The remaining 3.2 million MWh were sold in non-*Green-e Energy* certified transactions, most likely to utilities and electric service providers, power marketers, or retail customers.

Table 13. Total Retail Sales of Green-e Energy Certified Renewable Energy, 2008 and 2009 (Thousands of MWh)

	Residential		Commercial		Total Retail	
Year	2008	2009	2008	2009	2008	2009
RECs	50	40	10,490	15,653	10,540	15,693
Green Pricing	1,413	1,552	753	1,003	2,166	2,555
Competitive Electricity	171	224	170	188	341	411
Total	1,634	1,816	11,413	16,843	13,047	18,659

Note: Totals may not add due to rounding.
Source: Terada 2010.

Removing the instances of renewable energy certified by *Green-e Energy* at both the wholesale and retail levels, *Green-e Energy* certified sales of 21.9 million unique MWh in 2009. This is an increase of 26% from 2008. Assuming that all kilowatt-hours certified at the wholesale level were ultimately sold in retail voluntary sales, 73% of the total megawatt-hours sold in the retail voluntary market in 2009 were involved in a *Green-e Energy* certified transaction at some point in their chain of custody.

REC AND COMPETITIVE MARKET CUSTOMER PARTICIPATION

Participation in REC and competitive market programs nearly doubled, primarily due to new competitive offerings in Texas.

Based on data received from green power marketers, we estimate that more than 840,000 retail customers were buying green power from competitive suppliers or as unbundled RECs at the end of 2009 (see Table 14). This number includes about 130,000 participants in utility/marketer programs available in competitive markets. It is a particular challenge to obtain data about the competitive market, so it is likely that these figures underestimate the number of participants in competitive market programs.

The Texas market has seen dramatic growth in the number of green power offerings and participants in recent years.

The number of green power offerings in Texas has increased from 4 in November 2005 to 50 as of February 2010 (see Figure 7) (Power to Choose 2010). Texas saw the number of green power customers increase by 45%, from 142,000 customers in 2007 to 206,000 customers in 2008 (see Appendix B).[26] In 2009, participation in the Texas competitive market was likely more than 500,000; because NREL does not collect marketer data on a stateby-state level, the exact number of participants in the Texas market cannot be determined until EIA releases its customer data for 2009.

Gains in participation in Texas have been tempered by losses in some states, where marketers have struggled to provide electricity service to consumers amidst adverse market conditions and increasing costs. During 2008, EIA data show a slight decline in the number of green power customers in Pennsylvania and Virginia but slight gains in Maryland and D.C. (see Appendix B).

The increasing number of suppliers in Texas has been accompanied by increasing growth in voluntary retirements of RECs in Texas.

Voluntary REC retirements in Texas, including those by competitive marketers and utility green power programs, increased by 22.7% between 2008 and 2009, from 7.3 million MWh to 8.9 million MWh (ERCOT 2009).[27]

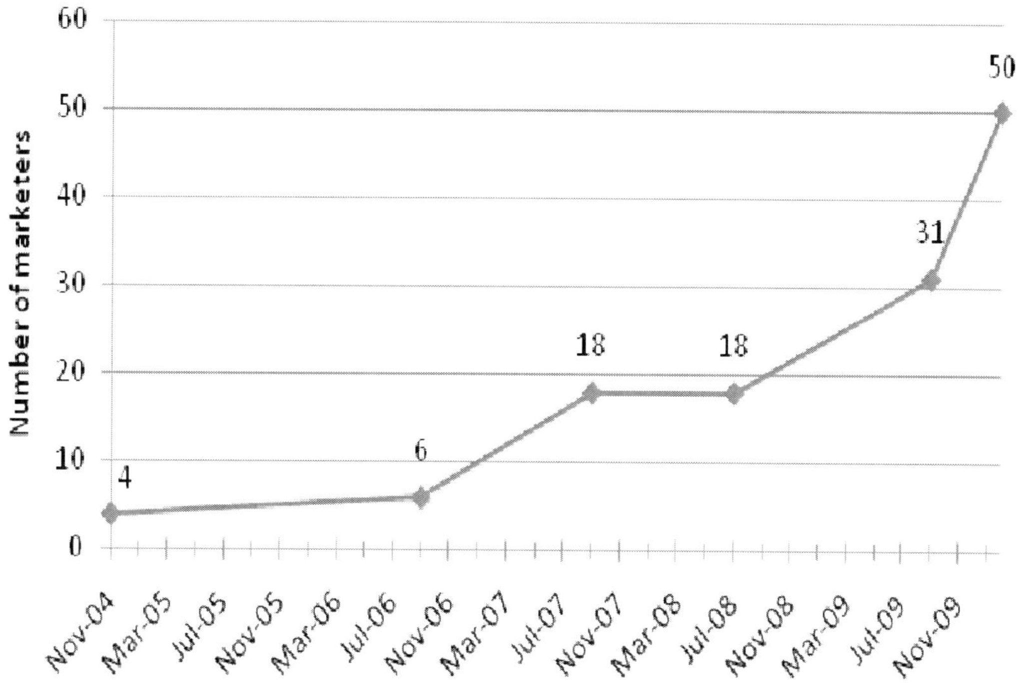

Figure 7. Texas green power product offers, 2004–2009.

A voluntary retirement occurs when a REC is used for voluntary purposes and will no longer be traded or claimed. Nationally, participation in utility/marketer partnership programs in competitive markets doubled between 2005 and 2008, although growth has slowed in the last two years.

In 2009, customer growth was similar to that of 2008, at 6%, while total sales declined by 7% from 2008 to 2009 in this sector. The decline in sales was prominent in two utility/marketer programs, which saw declines ranging from 21% to 47%. Figure 8 shows changes in both sales and customer participation in utility/marketer programs in competitive markets.

In competitive markets, the vast majority of customers buying green power are residential customers.

Of the approximately 840,000 retail green power customers in competitive markets, less than 2% purchase REC-only products.

The number of REC-only buyers increased from approximately 13,000 to 30,000 customers in 2008, showing some increase in traction with residential consumers—however, this trend reversed in 2009, when fewer than 20,000 customers purchased RECs.

While most of the REC buyers are residential customers, the majority of REC sales on a kilowatt-hour basis are made to nonresidential customers due to the much larger purchase sizes.

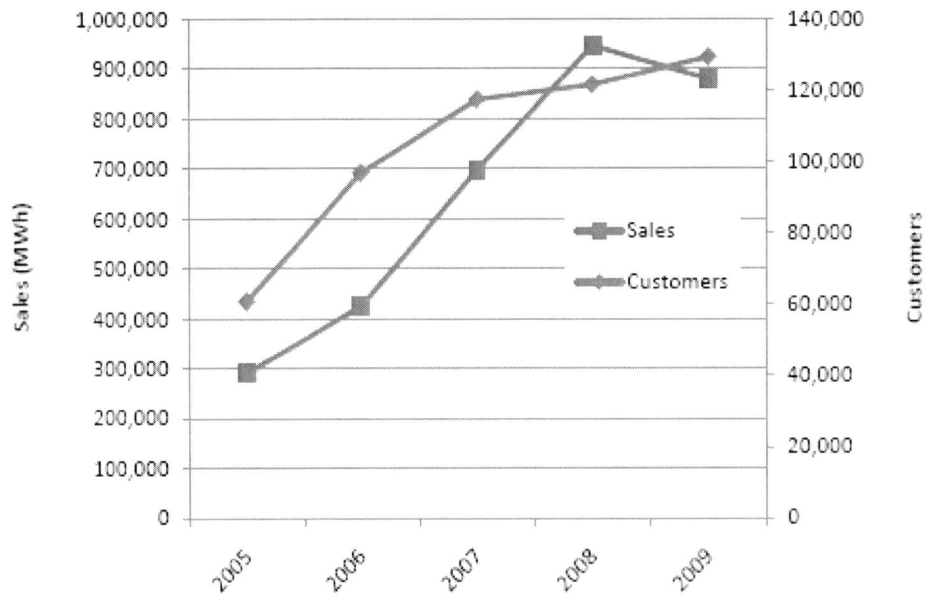

Figure 8. Changes in retail sales and customer participation for utility/marketer partnerships in competitive markets, 2005–2009.

Table 14. Estimated Cumulative Number of Customers Buying RECs or Green Power from Competitive Marketers, 2003–2009

	2003	2004	2005	2006	2007	2008	2009
Competitive Markets	- 170,000	< 140,000	> 180,000	- 210,000	- 300,000	- 390,000	- 830,000
RECs*	< 10,000	< 10,000	< 10,000	- 10,000	- 13,000	- 30,000	< 20,000
Total	- 180,000	< 150,000	- 190,000	- 220,000	> 310,000	- 425,000	< 850,000
% Change	13%	-17%	27%	16%	37%	37%	98%

*Includes only end-use customers purchasing RECs separate from electricity.

Note: Totals may not add due to rounding.

REC and Competitive Market Green Power Sales

Retail REC sales increased by 20% to 18.6 million MWh; competitive market sales increased 18% to 6.2 million MWh.

An estimated 24.8 million MWh of renewable energy was sold to retail customers by competitive green power and REC marketers in 2009 (Table 15). This figure includes renewable energy from both pre-existing and new sources. Due to the challenges of obtaining data from competitive marketers, it is likely that the sales figures for the competitive market are underestimated.

Retail REC sales increased by 20%, reaching 18.6 million MWh in 2009. Most of the growth in REC-only sales is attributable to the nonresidential sector.

An estimated 6.1 million MWh were sold as a bundled green power product in competitive electricity markets—an 18% increase from 2008. While there was a large increase in participation in competitive market programs, there was not a similarly sized increase in sales due to the fact that most of the growth in participation came from one marketer offering a product that was 25% green power.

2009 was a mixed year for both REC marketers and competitive marketers; some saw large gains in sales, while others saw sales remain flat or even down compared to 2008.

The competitive-market sales figure includes renewable energy sales through default utility/marketer programs or individual utility/marketer partnerships in competitive markets, which amounted to approximately 880 thousand MWh in 2009, a 7% decrease from 2008 (see Figure 8). The losses came primarily from two programs in the Northeast.

Table 15. Retail Sales of Renewable Energy in Competitive Markets and RECs, 2004–2009 (Thousands of kWh)

	2004	2005	2006	2007	2008	2009
Competitive Markets						
Residential	2,140	1,330	1,000	1,800	3,591	4,307
Nonresidential	510	820	710	1,400	1,669	1,879
Subtotal	2,650	2,150	1,720[a]	3,200	5,260[c]	6,186
% Change	40%	-19%	-20%[a]	88%[a]	64%[c]	18%
% Residential	81%	62%	59%	56%	68%	70%
Unbundled RECs[b]						
Residential	40	40	110	60	200	41
Nonresidential	1,690	3,840	6,700	10,500	15,400	18,619
Subtotal	1,720	3,890	6,810	10,500	15,600	18,660
% Change	160%	126%	75%	55%	49%	20%
% Residential	2%	1%	2%	1%	1%	0.2%
Total Sales	4,370	6,040	8,530	13,800	20,860	24,846
% Change	71%	38%	41%	62%	51%	19%

[a] 2006 are likely underestimated because of data gaps.

[b] Includes only RECs sold to end-use customers separate from electricity.

[c] 2008 competitive market sales were revised in this report to reflect data published in 2010 Note: Totals may not add due to rounding.

Table 15 also delineates green power sales by customer segment. In 2009, residential customers represented 70% of green power sales in competitive markets. In contrast, nonresidential customers represented nearly all unbundled REC sales. Generally, nonresidential customers find REC-only products attractive because of their flexibility and the greater potential for cost savings because they can be sourced from renewable energy projects in more favorable resource locations; also, the electricity does not have to be delivered directly to the customer, which lowers transaction costs. For commercial and institutional customers that operate facilities in multiple locations across the country, RECs may also provide a more efficient green power sourcing solution than working with utilities in each individual utility territory.[28] On the other hand, residential customers may not be aware that RECs are available or may not understand what they convey.

In 2009, renewable energy sold in competitive markets or as unbundled RECs represented an equivalent renewable energy capacity of about 7,710 MW, with more than 6,390 MW of this total coming from new renewable energy resources (see Table 16). 29,30 This is up from 5,800 MW of equivalent capacity and 4,900 MW of new capacity in 2008. Equivalent figures for 2007 are 3,700 MW and 3,000 MW, respectively.

Table 16. Renewable Energy Sources Supplying Competitive and REC Markets, 2009

	Biomass/ Landfill Gas	Geo-thermal	Hydro	Solar	Wind	Unknown	Total
Sales (thousands of MWh)	2,391	48	2,912	28	17,683	1,783	24,846
% of Total Sales	10%	0.19%	12%	0.11%	71%	7%	100%
Total MW	320	10	830	20	6,120	410	7,710
MW New Renewable Energy	260	10	420	20	5,680	--	6,390

Note: Information on new content is unavailable in some instances.

Note: Totals may not add due to rounding.

THE VOLUNTARY CARBON OFFSETS MARKET

Green power markets are affected by other related markets, such as the emerging US market for GHG offsets. Since green power and GHG offset offerings have converged in recent years, this section addresses GHG offsets sourced from renewable energy supplies. A GHG offset (sometimes referred to as a carbon offset) is a tradable commodity representing a unit of GHG emissions reduction or avoidance—typically, one metric ton of CO2e. Corporations and individuals are buying these products to "offset" their own emissions, such as those associated with energy used for heating, product manufacturing processes, automobile use, and air travel.

GHG offsets can be derived from a variety of project types that reduce or avoid GHG emissions, which use diverse methods for measuring these reductions. Examples of GHG reduction projects include renewable electricity generation, energy efficiency measures, methane capture at landfill sites, soil carbon sequestration, and forestry projects. Developers of these project types can sell GHG offsets to consumers or businesses to help finance their projects. For GHG offsets sourced from renewable energy generation projects, the equivalent emissions reduction of replacing conventional generation with renewable generation must be calculated. More than 25 companies offer offset products derived, at least in part, from renewable energy generation projects.[31]

Offsets sourced from renewable energy differ from green power in that they are sold in metric tons of CO2e, while RECs and other forms of green power are sold in megawatt-hours. In addition, certification standards for offsets differ from those for renewable energy sold as green power. Generally, offsets must demonstrate additionality, meaning that the emissions reductions are additional to what would have occurred anyway (or under business-as-usual). Retail customers typically purchase green power or RECs equivalent to a portion or all of their electricity consumption. In contrast, retail customers buying GHG offsets generally purchase metric tons of CO2e to match their carbon emissions. There is overlap in the sense that many green power purchasers are motivated to buy green power for their electricity consumption out of concern about climate change and to address their electricity-related GHG emissions. Currently, renewable energy generators can provide either a GHG offset (metric tons of CO2) or a megawatt-hour of green power—however; there are double-counting concerns if the same kilowatt-hour is sold as both an offset and a REC. Certifiers generally do not allow this type of double counting.

Eight out of approximately 25 GHG offset providers that offer products at least partially sourced from US-based renewable generation reported 2009 offset sales to NREL. Additional data on offset sales were obtained from the Center for Resource Solutions' Green-e Climate program.[32]

The carbon offsets sourced from renewable energy totaled more than 339,000 metric tons of CO2e, which is equivalent to about 486,000 MWh of renewable energy generation.33 This represents an increase of approximately 39% from 2008 when carbon offsets sources from renewable energy totaled nearly 246,000 metric tons of CO2e, or about 343,000 MWh of renewable energy generation.

Table 17. GHG Offsets Sourced from US-based Renewable Energy Sources, 2008–2009

	Carbon Offset Sales (Metric Tons CO_{2e})		Carbon Offset Sales (MWh equivalent)	
	2008	2009	2008	2009
Residential	31,200	45,400	43,500	67,800
Nonresidential	214,700	293,800	299,000	417,900
Total	245,900	339,200	342,500	485,700

Several independent certifiers have created standards for verifying GHG reductions to ensure that they are real, measurable, and beyond business-as-usual and any regulatory requirement. They also establish ownership of the actual emission reductions so that multiple parties do not claim the carbon reduction. GHG offset providers responding to the NREL questionnaire reported that some, if not all, of their offsets were verified by the following organizations: Center for Resource Solutions, Environmental Resources Trust,34 or the Chicago Climate Exchange (CCX).[35]

Voluntary Green Power Market Trends and Issues

As the voluntary green power market continues to grow, a few trends and issues have surfaced. This section highlights trends in REC prices in both the compliance and voluntary markets and discusses the current availability of data on REC prices and quantities transacted in the market and the general lack of price transparency.

The section concludes with a description of the treatment of renewable energy purchases in GHG inventories.

REC Prices

This section provides an overview of wholesale REC prices in voluntary and compliance markets in recent years based on indicative data available from brokers and third-party data providers. With a few exceptions, there is little price transparency in REC markets. Most transactions are conducted as bilateral contracts between parties, and prices are not reported. In addition, prices can vary widely by region. Therefore, data presented here are only indicative and should be used with caution.

In general, REC values depend on a number of factors, including the technology, the vintage (year in which it was generated), the volume purchased, the region in which the generator is located, whether they are eligible for certification, and whether the RECs are bought to meet compliance obligations or serve voluntary retail consumers. Natural gas prices can also affect the cost competitiveness of renewable energy generation, which is reflected in REC prices.

Compliance Markets for RECs

The region from which RECs are sourced is particularly important because often there are regional differences in renewable energy resource quality (e.g., wind speed) and electricity prices that determine the cost-effectiveness of the renewable generation. In addition, the supply and demand of RECs often varies regionally.

In regions where there have been shortages of renewable energy to meet RPS requirements, compliance REC prices have reached or come close to levels for alternative compliance payment (ACP) of \$50–\$55/MWh; whereas, in other states or regions, compliance RECs have sold for less than \$5/MWh. Figure 9 shows the wide variation in compliance REC prices among states for which data are available.

Solar RECs (SRECs) have higher value than RECs from other resource types in both compliance and voluntary markets. This is true for a number of reasons: 1) 16 states and D.C. have specific provisions to encourage solar or customer-sited generation (DSIRE 2010); 2) the penalty price for non-compliance is often set higher for solar/distributed generation tiers than for standard RPS compliance; and 3) SRECs can be desirable in the voluntary market where customers may be willing to pay more for solar, which costs more than other types of renewable energy.

Recently, PJM-GATS, the REC tracking system that covers the PJM regional transmission organization territory, began publishing the solar weighted average price for SRECs. The data date back to November 2008 and are updated on a monthly basis for SRECs traded in New Jersey, D.C., Ohio, Pennsylvania, Maryland, Delaware, and Virginia.[36] SRECs in New Jersey continue to trade at the highest levels, in the \$400–\$650 range, while SRECs from other regions trade in the \$200–\$500 range. While historic data availability is limited, several price points are indicative of the higher market price for SRECs in compliance markets in 2009 (see Figure 10).

In 2009, New Jersey also saw most of the SREC trading volume, at 85%, while Delaware, Maryland, and Pennsylvania saw smaller volumes, and Ohio, Virginia, and D.C., saw little to no volume (see Figure 11).

Voluntary Markets for RECs

While compliance RECs generally must be sourced from within some geographic region to be eligible for RPS compliance, voluntary RECs can be sourced either regionally or nationally. Most utility green pricing programs or marketers selling bundled electricity and REC products source their products from local or regional resources, with some exceptions.

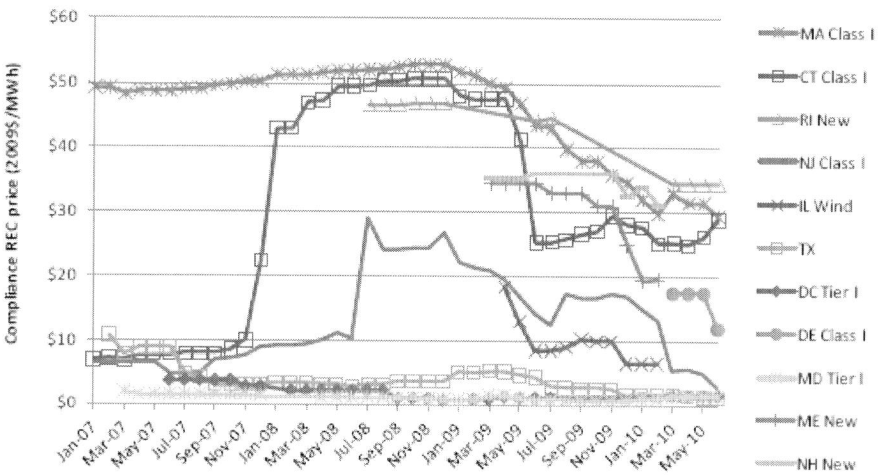

Sources: Evolution Markets (2007) and Spectron Group (2010).

Note: Plotted values are the last trade (if available) or the mid-point of bid and offer prices for the current or nearest compliance year for various state compliance RECs.

Figure 9. Compliance market (primary tier) REC prices, January 2007 to June 2010.

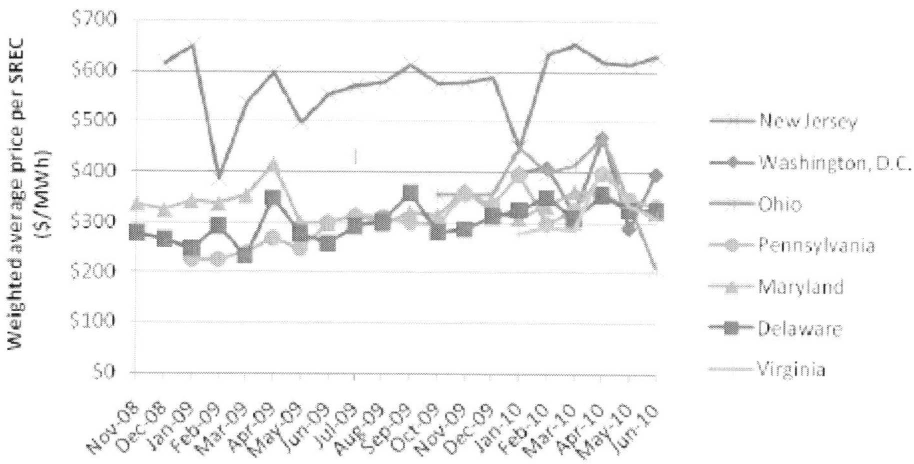

Source: PJM-GATS.

Figure 10. Compliance market SREC weighted average price, November 2008 to June 2010.

Buyers of nationally sourced voluntary RECs are often large corporations that have facilities in multiple locations across the country. In voluntary markets, RECs that are sourced locally (within the region) may have to compete with RPS demand or be subject to regional resource limitations. Therefore, regionally sourced voluntary RECs often sell at a premium to nationally sourced voluntary RECs, which are often derived from the most cost-effective renewable resources. As shown in Figure 12, wholesale RECs used in voluntary markets have generally traded in the range of $1/MWh to $10/MWh based on available indicative data.

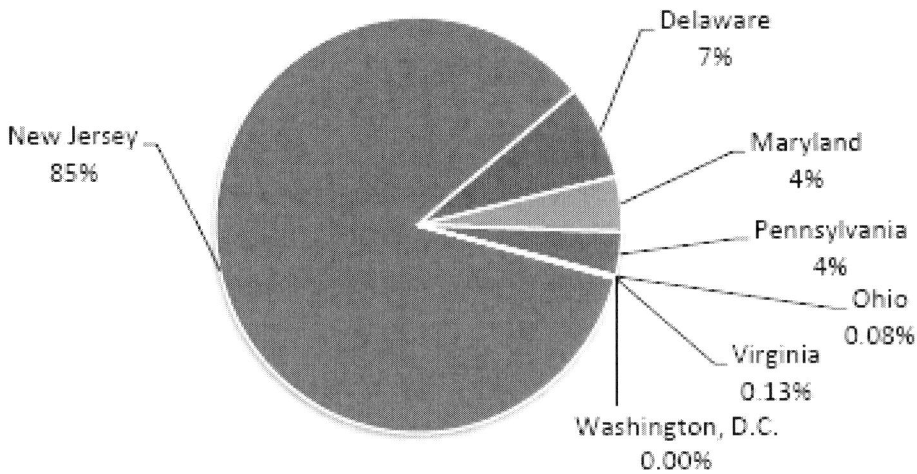

Source: PJM GATS.

Figure 11. State percent of annual SREC trading volumes, 2009.

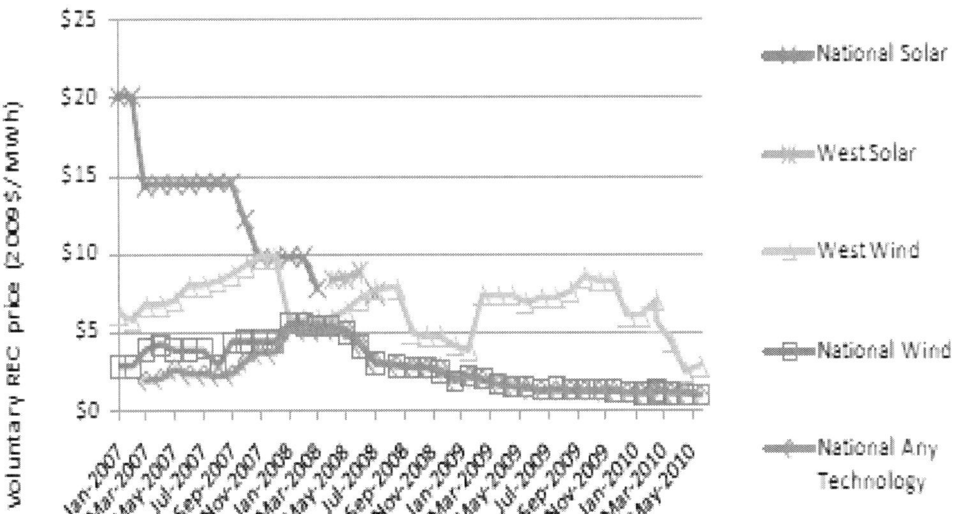

Sources: Evolution Markets (2007), Spectron Group (2010).

Figure 12. Voluntary REC prices, January 2007 to May 2010.

Table 18 presents wholesale voluntary REC prices for wind and for any renewable energy technology located nationally, as well as wind from within the Western Electric Coordinating Council (WECC). In 2009, prices paid for nationally sourced voluntary RECs from any technology ranged from about $0.88/MWh to $3.00/MWh. Nationally sourced voluntary wind REC prices were comparable to nationally sourced voluntary RECs for any technology, while wind from WECC netted higher prices on average. Prices differ not only by the technology and location but also by the vintage. Voluntary RECs sold in a given year can only be *Green-e Energy* certified if the renewable energy with which they are associated is generated in the calendar year in which the product is sold, the first three months of the following calendar year, or the last six months of the prior calendar year (CRS 2008). Table 18 shows price ranges for different vintages based on bids and offers in 2009 (ranges are based on the midpoint between bid and offer prices). In 2009, voluntary RECs from the 2008 vintage year were cheaper than current vintage RECs. In 2010, vintage year voluntary RECs were slightly more expensive than 2009 RECs, with the exception of WECC Wind.

REC Price Transparency and Quantity Information

Many renewable energy projects sell power and RECs bundled together in a single product. In such transactions, usually in the form of power purchase agreements (PPAs), there are no explicit REC prices.

Table 18. Range of Voluntary REC Prices in 2009 for Different Vintages ($/MWh)

Technology Type	2008	2009	2010
National Any	$0.88–$1.35	$1.00–$2.75	$1.35–$3.00
Technology			
National Wind	$0.88–$1.40	$1.03–$2.75	$1.35–$3.00
WECC Wind	$1.15–$5.25	$5.50–$8.75	$1.53–$2.08

Source: Spectron Group.

REC price information, therefore, is available only for RECs that are transacted separately, or unbundled, from the underlying power. REC prices, in both the voluntary and compliance markets, can be difficult to determine without the assistance of a broker, and even then, available information only indicates the transactions made by one broker. A few sources offer publicly available data on SREC pricing. PJM-GATS, the state of New Jersey, and brokers at SRECTrade, Flett Exchange, and PJM EnviroTrade all provide various forms of SREC pricing publicly. In addition to these sources, a few jurisdictions (Maryland, Pennsylvania, and D.C.) require that REC prices be disclosed.

In recent years, more SREC pricing data has become publicly available. SRECTrade, Flett Exchange, and PJM EnviroTrade post SREC data on their Web sites. SRECTrade[37] was founded in 2007 and launched its online platform in August 2008. The founders developed the platform because of the "lack of a transparent, fair and public marketplace that facilitated the sale of SRECs in a cost effective way" (SRECTrade 2010). SRECTrade runs auctions for SRECs from Delaware, Massachusetts, Maryland, New Jersey, Ohio, Pennsylvania, and D.C.

Price information is available on a monthly basis for each state. Historical price information dates back to September 2009 when the exchange began auctioning New Jersey SRECs.

Flett Exchange [38] was also founded on the principle that markets should be more transparent. The exchange began in 2006. Flett Exchange posts results from its REC auctions on its Web site and in a monthly newsletter. In addition to compliance SRECs, Flett Exchange brokers voluntary RECs and also posts price and volume information on its voluntary transactions.

PJM EnviroTrade[39] is a subsidiary of PJM Interconnection, the regional transmission organization that coordinates electricity markets in all or parts of 13 states in the mid-Atlantic and Midwest.[40] PJM EnviroTrade began a monthly auction of SRECs in the summer of 2010. In June 2010, PJM EnviroTrade began posting monthly updates of its auction results, listed in volume, offer price range, and bid price range, by state.

In New Jersey, SREC data is posted on the NJ Clean Energy [41] Web site on a monthly basis, with a delay of one to two months. The data show low and high SREC prices and the weighted average cost. Historical data are available dating back to August 2004. In the current reporting year (June 1, 2009–May 31, 2010), the New Jersey SREC pricing data is sourced from trades on the PJM-GATS trading platform.

PJM-GATS, the generation attribute tracking system for PJM, provides the high, low, and weighted average SREC price by state, dating back to November 2008. Data is available from Delaware, Maryland, North Carolina, New Jersey, Ohio, Pennsylvania, Virginia, and D.C. Data is posted on a monthly basis and can be queried by state. In addition to posting SREC prices, PJM-GATS provides data to Pennsylvania, New Jersey, Maryland, and D.C., on the price paid for RECs that are retired for RPS compliance.

While there have been some emerging data sources for SREC prices, REC price transparency remains limited. Pennsylvania posts annual REC ("Alternative Energy Credit") pricing online[42], including the weighted average price and price range for each tier of resources. However, because pricing is delayed and only reported on an annual basis, it provides only an historic price perspective. In the 2008–2009 compliance year, pricing for solar PV ranged from $225 to $690, with a weighted average price of $260, while Tier I pricing ranged from $0.50 to $23.00, with a weighted average price of $3.65.

Washington, D.C., requires utilities to submit price information in their RPS filings; however, the information is not aggregated and published in any form. While this compliance information is still helpful to the PUC, it does not do much to foster greater market price transparency because it is unlikely that individuals looking for REC price information would spend the time to search through individual case filings on the PUC Web site for price information.

REC Quantity Information

While there are sensitivities around the price paid for RECs, there are few concerns about increasing the transparency of the quantity of RECs retired for compliance or voluntary purposes. The Texas PUC has encouraged public access to REC market data by requiring ERCOT to annually report the aggregate quantity of RECs retired for voluntary and compliance purposes. In the current reporting year, confidentiality is ensured to account holders, which may be retiring compliance and/or voluntary RECs, but after one year, confidentiality is expired, and ERCOT documents how many RECs were retired by each account holder.[43]

WREGIS, PJM-GATS, MRETS, and NC-RETS all track voluntary retirements, but none of them currently make the information publicly available. Several REC tracking systems have indicated that they are willing to share or make public the quantity of RECs retired in a given year, subject to approval by their respective governing boards. PJM-GATS and MRETs are moving forward with providing data publicly (Schuyler 2010; Gower 2010).

Treatment of Renewable Energy Purchases in GHG Inventories

Because many organizations are purchasing renewable energy or RECs as part of their comprehensive GHG strategies, questions have arisen regarding how best to treat RECs and green power purchases under carbon accounting methodologies. Leading GHG inventories allow participants to account for renewable energy purchases, although there are some differences in methods used, particularly with respect to calculating and crediting the emissions benefits.

Renewable energy purchases are generally matched with purchased electricity (i.e., Scope 2 emissions), which are considered indirect emissions because they are not under the direct control of the facility (see Figure 13). However, there are some differences in the methods in which they are reported and how the adjustment to Scope 2 emissions is calculated. To give an indication of the differences of methodologies in use, we briefly summarize the methods that leading public GHG inventories use to account for renewable energy purchases.

EPA Climate Leaders Program

The US EPA Climate Leaders Program[44] is a voluntary program under which companies develop long-term climate change goals and develop a corporate-wide GHG inventory to track progress. The program uses an inventory protocol developed by the World Resource Institute and the World Business Council for Sustainable Development. The EPA Climate Leaders Program recognizes internal reductions including efficiency, on-site renewable energy, project-based offsets, and green power purchases (including bundled renewable electricity as well as unbundled RECs).[45] In September, 2010, EPA announced that it would be phasing out the Climate Leaders Program over the next year, and encouraged its participating companies to join another state or non-governmental program. However, it is still useful to examine how the program has addressed GHG accounting of green power, as it provides perspective on how US EPA has addressed these issues to date.

For on-site renewable generation, no adjustment to the GHG inventory is required as it will already be reflected in GHG inventory. Yet if the partner sells RECs from an on-site power generation facility, then the renewable energy generation cannot be counted toward GHG reductions; instead, the indirect emissions associated with the electricity equivalent to the RECs sold from the on-site facility must be reported.

Renewable energy purchases and RECs can be used to adjust GHG inventory emissions if the RECs meet all the resource eligibility and additionality requirements specified by EPA, including being additional to regulatory mandates. Partners are able to use green power purchases to reduce GHG emissions associated with their purchased power (Scope 2 emissions).

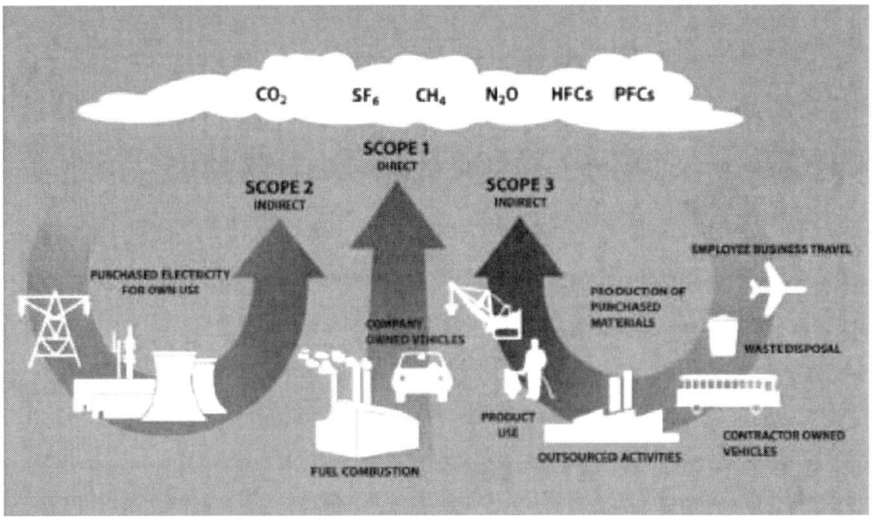

Source: WRI and WBCSD (2004).

Figure 13. Overview of scopes and emissions.

Scope 2 emissions can be reduced by the product of the avoided emission rate of the renewable energy generator and the amount of green power purchased.[46]

The Climate Registry

The Climate Registry[47] is a nonprofit organization that provides standards for businesses and governments to calculate, verify, and publicly report their North American carbon footprints in a single, unified registry. The Climate Registry supports both mandatory and voluntary reporting programs and is guided by a Board of Directors comprised of 41 US states and Washington, D.C., 13 Canadian provinces and territories, 6 Mexican states, and 4 Native Sovereign Nations.

The Climate Registry was established in 2007 as an extension of the California Climate Action Registry (CCAR), which has been helping companies in California voluntarily report direct and indirect emissions from their operations in California since 2001.

The Climate Registry's General Reporting Protocol (GRP) gives entities the option to report renewable energy purchases as supplemental information to their Scope 2 emissions data.

The GRP called for entities to report their Scope 2 emissions based on the system average emissions mix (without a direct adjustment for renewable energy purchases, as in the Climate Leaders program), but entities could calculate the benefits of green power purchases and provide that as supplemental information.[48]

The Climate Registry is currently conducting a pilot program in which renewable energy purchases (including RECs) can be used to adjust Scope 2 emissions (Foran 2010).

The program, called the Climate Registered Program, was announced in December 2009 and is designed to recognize leading organizations that meet GHG emissions reductions goals consistent with the program's silver, gold, and platinum leadership levels.

Under the program, participating organizations can use renewable energy purchases (including RECs) or offsets to meet a portion of their GHG reduction goals. The green power purchases can be reported as an adjustment to Scope 2 (purchased electricity) emissions by matching the megawatt-hours of green power or RECs with the megawatt-hours of electricity purchases, starting with the electricity purchases in the cleanest region to yield a conservative estimate of emissions benefits. For example, a company that purchases electricity for facilities in California and Illinois would first credit the green power purchases against the California electricity consumption, which has a lower regional GHG emissions rate.

In contrast to the Climate Leader's methodology, this method does not rely on estimating the avoided emissions based on the location of the renewable energy generator supplying the green power (which in some cases is not known) but rather applying the zero-emissions benefits of the renewable energy to the organization's purchased electricity, whose estimated emissions are based on the regional grid mix. Participating organizations will report two Scope 2 emissions levels: a) an unadjusted estimate of emissions based on the local emissions of the purchased electricity (using EPA eGRID regional emissions factors) and b) an adjusted estimate of emissions including any Registry-accepted utility-specific emission factors and any renewable energy purchases or offsets (Foran 2010).

The Climate Registry plans to evaluate the pilot program after about six months and anticipates receiving public comments on the program and adjustment methods employed. Thus, the calculation methodologies and program structure could be modified going forward based on stakeholder input.

The Climate Registry has also developed an industry-specific protocol for the power sector that incorporates a framework to deal with RECs sold by power companies that own renewable energy facilities or procure RECs for their portfolio.[49] Under the Climate Registry's electric sector protocol, utilities have the option to report sales and purchases of RECs sold from renewable energy facilities that they own. Utilities following the optional protocol are required to estimate system average emissions for those renewable energy facilities in which RECs are not retained. This ensures that the emissions benefits are not double counted by the utility and the renewable energy purchaser.

Local Government Operations Protocol

ICLEI – Local Governments for Sustainability (ICLEI) is an association of city and county governments interested in making their communities more sustainable. ICLEI assists local governments in their efforts to reduce GHG emissions, providing tools and methods to measure emissions. ICLEI developed the Local Government Operations Protocol[50] ("Protocol") in partnership with the California Air Resources Board (CARB) and CCAR and in collaboration with The Climate Registry.

The Protocol is designed specifically for use by local governments throughout the United States. CARB encourages California's local governments to use the Protocol to annually inventory and report their GHG emissions so that reductions made by local governments are transparent, consistent, and accurate.

For local governments that purchase renewable energy, either through their electric utility or an independent power provider, the Protocol does not allow deductions from Scope 2 emissions. The Protocol made this determination because it considers this purchase to be already accounted for in the region's emissions rate, or eGRID factor. [51]

However, local governments are encouraged to report renewable energy purchases as supplemental information in their GHG inventories. For local governments generating on-site renewable energy, the Protocol does not require any adjustment because the generation will decrease the grid electricity that is purchased, therefore decreasing Scope 2 emissions.

Draft Federal Guidance on Executive Order 13514

Executive Order 13514 was signed on October 5, 2009, requiring the federal government to make reductions of GHG emissions a priority for federal agencies.

Each federal agency is directed to establish a percentage reduction target relative to a fiscal year 2008 baseline and complete a GHG inventory of fiscal year 2010 emissions by January 31, 2011. The Federal GHG Accounting and Reporting Guidance ("Guidance") is being developed by the Department of Energy's Federal Energy Management Program (FEMP), in coordination with other federal agencies. The draft Guidance was released for public comment on July 14, 2010.[52]

According to the draft Guidance, federal agencies may reduce their Scope 2 emissions when purchasing renewable energy or RECs. Purchases of electricity from renewable energy generators must include ownership of the REC in order to qualify as renewable. The draft Guidance requires GHG emissions adjustments for renewable energy purchases to be calculated based on the non-baseload eGRID emission rate of the region where the renewable generator is located.

On-site renewable generation, when the associated RECs are owned by the agency, will reduce Scope 1 emissions if the renewable generation is displacing on-site non-renewable generation. If the on-site renewable generation is displacing purchased electricity, Scope 2 emissions will be reduced due to the decreased use of purchased electricity. If an agency does not own the RECs from their on-site renewable generation, the agency must adjust its Scope 1 and Scope 2 emissions by reporting emissions for the electricity associated with the REC as if it were conventional electricity by using the non-baseload emissions factor for the eGRID sub-region of the on-site renewable generation system.

Conclusions and Observations

The green power market continues to exhibit strong growth and provide an important demand-driven stimulus for renewable energy development. Green power markets provide an additional revenue stream for renewable energy projects and raise consumer awareness of the benefits of renewable energy. Based on this review, we have identified the following market trends:

- In 2009, total retail sales of renewable energy in voluntary-purchase markets exceeded 30 billion kWh, representing a capacity equivalent of 9,500 MW of renewable energy, including 8,300 MW from "new" renewable energy sources.
- Wind energy provided 73.7% of total green power sales, followed by biomass energy sources including landfill gas (10.0%), hydropower (9.9%), geothermal (0.2%), solar (0.1%), and the remainder unknown (5.9%).
- Total market sales increased by 17% in 2009, dominated by REC sales, which are primarily to nonresidential consumers and increased by about 20% from 2008. REC markets now represent 62% of green power market sales, surpassing sales in competitive electricity markets and utility green pricing programs.
- Overall, the total number of customers purchasing green power increased by 44% in 2009, a higher rate than in previous years, with gains coming primarily from one

competitive offering in Texas. Utility green pricing program participants remained essentially flat in aggregate from 2007 to 2009, with some programs continuing to report customer losses in 2009, presumably due to the economic downturn.

- Utility green pricing programs in regulated electricity markets continued to grow on a sales basis but at a slower rate than in previous years, with sales increasing by about 7% in 2009. A relatively small number of utility programs continued to dominate sales and customer results. Utility premiums for green pricing have continued to fall due in part to the increased cost competitiveness of renewable energy with conventional generation.
- In 2009, nearly 340,000 metric tons of CO_2e avoided due to renewable energy facilities were marketed as offsets, an increase of approximately 39% from 2008. This is the equivalent of about 485,000 MWh of renewable energy generation.
- In 2009, sales to nonresidential customers continued to outpace those to residential customers, bringing the fraction of nonresidential sales to 76% of all green power sales on a kilowatt-hour basis. The continuing dominance of nonresidential sales is a departure from the early history of green power markets when most products and programs were oriented toward residential customers.

REC prices vary considerably, depending on a number of factors. In compliance regions where there have been shortages of renewable energy to meet RPS requirements, REC prices have reached or come close to levels for ACP of $50– $55/MWh; whereas, in other states or regions, compliance RECs have sold for less than $5/MWh.

Wholesale RECs used in voluntary markets have generally traded in the range of $1–$10/MWh.

APPENDIX A. LEADING PURCHASERS IN THE EPA GREEN POWER PARTNERSHIP

Table A-1. Top 25 Purchasers in the EPA Green Power Partnership Program, January 5, 2010

Rank	Company	Annual Green Power Usage (kWh)	GP % of Total Electricity Use	Green Power Resources
1	Intel Corporation	1,433,200,000	51%	Biogas, Biomass, Geothermal, Small Hydro, Solar, Wind
2	Kohl's Department Stores	1,367,376,000	100%	Biogas, Biomass, Small Hydro, Solar, Wind
3	PepsiCo	1,226,403,121	100%	Various
4	Whole Foods Market	790,459,000	105%	Solar, Wind
5	City of Houston, TX	438,000,000	34%	Wind
6	Dell, Inc.	431,058,000	129%	Biogas, Solar, Wind
7	The Pepsi Bottling Group, Inc.	426,239,848	100%	Various
8	Cisco Systems, Inc.	400,996,000	46%	Wind

Table A-1. (Continued)

Rank	Company	Annual Green Power Usage (kWh)	GP % of Total Electricity Use	Green Power Resources
9	Commonwealth of Pennsylvania	400,000,000	40%	Biomass, Wind
10	Johnson & Johnson	386,455,711	34%	Biogas, Biomass, Small Hydro, Solar, Wind
11	US Air Force	339,660,392	4%	Biogas, Biomass, Solar, Wind
12	City of Dallas, TX	333,659,840	40%	Wind
13	HSBC North America	314,013,000	98%	Wind
14	US Environmental Protection Agency	262,262,425	101%	Biogas, Biomass, Solar, Wind
15	Wal-Mart Stores, Inc./California and Texas Facilities	243,328,000	8%	Solar, Wind
16	Starbucks	237,000,000 46	25%	Wind
17	BNY Mellon	229,500,000	77%	Wind
18	City of Chicago, IL	214,635,000	20%	Biomass, Wind
19	Kimberly-Clark Corporation	192,730,000	7%	Biomass
20	University of Pennsylvania	192,727,000	46%	Wind
21	US Department of Energy	188,599,600	4%	Various
22	Los Angeles County Sanitation Districts	181,624,000	55%	Biogas
23	DuPont Company	180,075,000	4%	Biomass, Solar, Wind
24	Wells Fargo & Company	175,000,000	14%	Wind
25	Deutsche Bank	160,000,000	100%	Wind

Source: EPA Green Power Partnership, .http://www.epa.gov/greenpower/toplists/ top50.htm.

APPENDIX B. ESTIMATED US GREEN PRICING CUSTOMERS

Table B-1. Estimated US Green Pricing Customers by State and Customer Class, 2007 and 2008

State	Electric Industry Participants 2008[a]		Participating Customers		
			2008		2007
		Residential	Nonresidential	Total	Total
Alabama	25	1,786	30	1,816	585
Alaska	1	440	20	460	530
Arizona	6	4,222	123	4,345	9,285
Arkansas	2	25	0	25	-
California	13	80,178	3,432	83,610	58,676
Colorado	26	56,270	1,966	58,236	57,501
Connecticut	3	122	24	146	96
Delaware	9	11,193	1,260	12,453	8,914
D.C.	3	1,590	3,925	5,515	4,854
Florida	5	38,099	385	38,484	37,833

State	Electric Industry Participants 2008[a]	Participating Customers			
		2008			2007
		Residential	Nonresidential	Total	Total
Georgia	24	9,170	186	9,356	8,308
Hawaii	-	-	-	-	4,738
Idaho	6	4,935	192	5,127	4,817
Illinois	4	4,225	40	4,265	3,892
Indiana	14	6,111	97	6,208	4,299
Iowa	40	8,522	743	9,265	9,193
Kansas	1	1	-	1	1
Kentucky	24	3,026	32	3,058	1,338
Louisiana	2	357	38	395	-
Maine	2	2,003	218	2,221	2,494
Maryland	4	42,690	16,337	59,027	55,954
Massachusetts	8	9,738	474	10,212	6,155
Michigan	11	27,843	285	28,128	13,196
Minnesota	98	43,879	554	44,433	44,034
Mississippi	12	249	9	258	3
Missouri	20	4,283	55	4,338	1,439
Montana	11	538	26	564	995
Nebraska	4	7,585	61	7,646	6,891
Nevada	2	30	1	31	514
New Hampshire	1	0	1	1	1
New Jersey	4	1,945	323	2,268	441
New Mexico	11	3,129	300	3,429	21,273
New York	9	27,310	1,225	28,535	21,857
North Carolina	23	13,936	287	14,223	12,386
North Dakota	7	3,095	14	3,109	5,086
Ohio	13	3,625	130	3,755	1,789
Oklahoma	8	9,882	539	10,421	11,287
Oregon	24	109,656	3,442	113,098	100,595
Pennsylvania	5	36,742	812	37,554	39,099
Rhode Island	2	5,086	120	5,206	4,887
South Carolina	21	9,895	485	10,380	4,766
South Dakota	7	596	16	612	632
Tennessee	64	11,712	987	12,699	-
Texas	18	184,994	20,731	205,725	142,334
Utah	8	25,291	607	25,898	23,406
Vermont	2	4,535	257	4,792	4,517
Virginia	2	1,062	0	1,062	1,306
Washington	25	46,516	1,391	47,907	43,885
West Virginia	2	72	2	74	-
Wisconsin	60	45,889	2,229	48,118	36,344
Wyoming	8	4,206	300	4,506	13,225
Total	643	918,284	64,711	982,995	835,651

[a] Includes entities with green pricing programs in more than one state.

- = No data reported.

Note: Nonresidential may include some customers for whom no customer class is specified. Note: Totals may not add due to rounding.

Source: Energy Information Administration, "Green Pricing and Net Metering Programs, 2008." http://www.eia.gov/cneaf/solar.renewables/page/greenprice/ netmetering08.pdf. Accessed August 2010.

Table B-2. Estimated US Green Pricing Customers by Customer Class, 2002–2008

Year	Electric Industry Participants	Participating Customers		
		Residential	Nonresidential	Total
2002	212	688,069	23,481	711,550
2003	308	819,579	57,547	877,126
2004	403	864,794	63,539	928,333
2005	442	871,774	70,998	942,772
2006[a]	484	606,919	35,937	642,856
2007	591	773,391	62,260	835,651
2008	643	918,284	64,711	982,995

Source: Energy Information Administration, "Net Metering and Green Pricing Customers by End Use Sector, 2002 - 2008," http://www.eia.doe.gov/cneaf/ electricity/epa/epaxlfile7 5.pdf. Accessed January 2010.

Note: Nonresidential may include some customers for whom no customer class is specified.

[a] In 2006, the single largest provider of green pricing services in the country discontinued service in two States. More than 297,600 customers in green pricing programs reverted to standard service tariffs, predominantly in Ohio and Pennsylvania.

APPENDIX C. UTILITIES OFFERING GREEN PRICING PROGRAMS IN REGULATED MARKETS

Table C-1. Utilities Offering Green Pricing Programs in Regulated Markets, 2009

Investor-Owned Utilities	Electric Cooperatives	Municipal/Public Utilities	Muscatine Power and Water
AEP Appalachian Power	Alabama Electric Cooperative	City of Alameda	City of Naperville
Alliant Energy	Associated Electric Cooperative, Inc.	American Municipal Power-Ohio	City of New Smyrna Beach
AmerenUE	Bandera Electric Cooperative	Anaheim Public Utilities	Northern Wasco County PUD
Arizona Public Service	Basin Electric Power Cooperative	City of Ashland	Oklahoma Municipal Power Authority
Avista Utilities	Boone Electric Cooperative	Austin Energy	Omaha Public Power District
Central Vermont Public Service	Buckeye Power	Austin Utilities (MN)	Owatonna Public Utilities
Cheyenne Light, Fuel and Power Co.	Central Electric Cooperative	Benton County Public Utility District	Pacific County PUD
Connecticut Light and Power	Central Iowa Power Cooperative	City of Bowling Green	City of Palo Alto Utilities
Consumers Energy	Connexus Energy	Braintree Electric Light Department	Pasadena Water & Power
Dayton Power and Light	Corn Belt Power Cooperatives	Burbank Water and Power	Platte River Power Authority
Dominion North Carolina Power	Dairyland Power Cooperative	CPS Energy (San Antonio)	Roseville Electric
Dominion Virginia Power	Dakota Electric Association	Cedar Falls Utilities	Sacramento Municipal Utility District
DTE Energy	Delaware Electric Cooperative	Central MN Municipal Power Agency	Salt River Project
Duke Energy	Deseret Power	Chelan County Public Utility District	San Francisco Public Utilities Commission
El Paso Electric Company	Deseret Pow er/Mt. Wheeler Pow er Cooperative	Clallam County PUD	Santee Cooper
Entergy Gulf States	East Kentucky Power	Clark Public Utilities	Seattle City Light

Investor-Owned Utilities	Electric Cooperatives	Municipal/Public Utilities	Muscatine Power and Water
	Cooperative		
E.ON US	Electric Cooperatives of Arkansas	College Station Utilities (TX)	Shrewsbury Electric and Cable Operations
FirstEnergy	Farmers Electric Cooperative	Colorado Springs Utilities	Silicon Valley Power
Georgia Power	Flathead Electric Cooperative	Columbia River PUD	Snohomish County Public Utility District
Green Mountain Power	Georgia Electric Membership Corporation	Concord Municipal Light Plant	Southern Minnesota Municipal Power Agency
Gulf Power Company	Golden Valley Electric Association	Cowlitz PUD	City Utilities of Springfield (MO)
Hawaiian Electric Company	Great River Energy	Edmond Electric	Springfield Utility Board
Idaho Power Company	Gunnison County Electric Association	City of Eldridge (IA)	City of St. Charles
Indianapolis Power & Light Company	Holy Cross Energy	ElectriCities	City of St. George
Kansas City Power & Light	Hoosier Energy	Emerald People's Utility District	Tacoma Power
Kentucky Power Co.	Intermountain Rural Electric Association	Estes Park Light and Power	City of Tallahassee
Kentucky Utilities Company	KAMO Electric Cooperative	Eugene Water & Electric Board	Truckee Donner Public Utility District
Louisville Gas and Electric Company	Kauai Island Utility Cooperative (KIUC)	Fort Collins Utilities	Waverly Light and Power
Madison Gas and Electric	La Plata Electric Association	Gainesville Regional Utilities	WPPI Energy
MidAmerican Energy	Lower Colorado River Authority	Grant County PUD	
Minnesota Power	Lower Valley Energy	Grays Harbor PUD	Federal
NSTAR Electric	Midstate Electric Cooperative	Heartland Consumers Power District	Tennessee Valley Authority
Nevada Power	Minnkota Power Cooperative	Iowa Association of Municipal Utilities	
NorthWestern Energy	New-Mac Electric Cooperative	Keys Energy Services	
OG&E Electric Services	Orcas Power & Light	Lakeland Electric	
Otter Tail Power Company	Oregon Trail Electric Cooperative	Lansing Board of Water and Light	
PacifiCorp	Palmetto Electric Cooperative	Lenox Municipal Utilities	
Portland General Electric Company	Park Electric Cooperative	Lewis County PUD	
Progress Energy	Pedernales Electric Cooperative	Lincoln Electric System	
Public Service Company of NM	Peninsula Light Company	Lodi Utilities	
Puget Sound Energy	Power South Energy Cooperative	Longmont Power & Communications	
SCE&G	PNGC Power	Los Alamos County (NM)	
Tampa Electric Company	Rappahannock Electric Cooperative	Los Angeles Dept. of Water and Power	
Tucson Electric Power Company	Southern Montana Electric G&T Cooperative	Loveland Water & Power	
UniSource Energy Services	Tri-State Generation and Transmission Ass	Mason County PUD No. 3	
United Illuminating	Vigilante Electric Cooperative	Missouri Joint Municipal Electric Utility	

Table C-1. (Continued)

Investor-Owned Utilities	Electric Cooperatives	Municipal/Public Utilities	Muscatine Power and Water
Upper Peninsula Power Company	Wabash Valley Power Association	Missouri River Energy Services	
We Energies	Western Farmers Electric Cooperative	Moorhead Public Service	
Wisconsin Public Service Corporation	Yampa Valley Electric Association		
Xcel Energy			

Table C-2. Utility/Marketer Green Power Programs in Restructured Electricity Markets, 2009

State	Utility
Connecticut	Connecticut Light & Power United Illuminating
Maine	Kennebunk Light and Power District
Massachusetts	Massachusetts Electric (National Grid) Nantucket Electric (National Grid)
Michigan	Consumers Energy
New Jersey	Atlantic City Electric Public Service Electric & Gas Rockland Electric Jersey Central Power & Light Orange and Rockland Utilities
New York	Long Island Power Authority Energy East/NYSEG Niagara Mohawk (National Grid)
Pennsylvania	PECO Energy
Rhode Island	Narragansett Electric (National Grid)

Appendix D. Links to Utility Green Pricing Programs and REC and Competitive Market Green Power Offerings

Table of Utility Green Pricing Programs by State:
http://www.eere.energy.gov/greenpower/markets/pricing.shtml?page=1
REC Retail Products:
http://www.eere.energy.gov/greenpower/markets/certificates.shtml?page=1
Retail Green Power Product Offerings in States with Retail Competition: http://www.eere.energy.gov/greenpower/markets/marketing.shtml?page=1

Appendix E. Top 10 Utility Green Pricing Programs

Table E-1. Top 10 Green Pricing Program Renewable Energy Sales (as of December 2009)

Rank	Utility	Resources Used	Sales (kWh/year)	Sales (aMW)[a]
1	Austin Energy	Wind, Landfill Gas	764,895,830	87.3
2	Portland General Electric[b]	Wind, Biomass, Geothermal	740,880,487	84.6
3	PacifiCorp[cde]	Wind, Biomass, Landfill Gas, Solar	578,744,080	66.1
4	Sacramento Municipal Utility District[c]	Wind, Hydro, Biomass, Solar	377,535,530	43.1
5	Xcel Energy[cf]	Wind, Solar	374,296,375	42.7
6	Puget Sound Energy[cg]	Wind, Landfill Gas, Biomass, Small Hydro, Solar	303,046,167	34.6
7	Connecticut Light and Power/ United Illuminating	Wind, Hydro	197,458,734	22.5
8	National Grid[h]	Biomass, Wind, Small Hydro, Solar	174,536,130	19.9
9	Public Service Company of New Mexico	Wind	173,863,751	19.8
10	We Energiesc	Wind, Landfill Gas, Solar	173,217,802	19.8

[a] An "average megawatt" (aMW) is a measure of continuous capacity equivalent (i.e., operating at a 100% capacity factor).
[b] Marketed in partnership with Green Mountain Energy Company.
[c] Product is Green-e Energy (www.green-e.org) certified.
[d] Some Oregon products marketed in partnership with 3Degrees Group Inc.
[e] Includes Pacific Power and Rocky Mountain Power.
[f] Includes Northern States Power, Public Service Company of Colorado, and Southwestern Public Service.
[g] Residential product marketed in partnership with 3Degrees Group Inc.
[h] Includes Niagara Mohawk, Massachusetts Electric, Narragansett Electric, and Nantucket Electric.

Table E-2. Total Number of Customer Participants ants as of December 2009

Rank	Utility	Program(s)	Participants
1	Portland General Electric[a]	Clean Wind, Green Source, Renewable Future	72,812
2	PacifiCorp [bc]	Blue Sky Blockd, Blue Sky Usage[d] Blue Sky Habitat[d]	71,165
3	Xcel Energy[e]	WindSourced, Renewable Energy Trust	70,393
4	Sacramento Municipal Utility District	Greenergy[d]	50,250
5	PECO[f]	PECO WIND	34,491
6	Puget Sound Energy[cg]	Green Power Program[d]	25,789
7	National Grid[h]	GreenUp	22,888
8	Connecticut Light and Power/ United Illuminating	CTCleanEnergyOptions	22,336
9	We Energies	Energy for Tomorrow[d]	20,927
10	Iberdrola USA: NYSEG and RG&E[f]	Catch the Wind	20,386

[a] Marketed in partnership with Green Mountain Energy Company.
[b] Includes Pacific Power and Rocky Mountain Power.
[c] Some Oregon products marketed in partnership with 3Degrees Group Inc.
[d] Product is Green-e Energy certified.
[e] Includes Northern States Power, Public Service Company of Colorado, and Southwestern Public Service.
[f] Marketed in partnership with Community Energy Inc.
[g] Residential product marketed in partnership with 3Degrees Group Inc.
[h] Includes Niagara Mohawk, Massachusetts Electric, Narragansett Electric, and Nantucket Electric.

Table E-3. Customer Participation Rate (as of December 2009)

Rank	Utility	Program(s)	Customer Participation Rate	Program Start Year
1	City of Palo Alto Utilities[a]	Palo Alto Green[b]	20.8%	2003
2	Portland General Electric[c]	Clean Wind, Green Source, Renewable Future	10.2%	2002
3	Madison Gas and Electric	Green Power Tomorrow	9.6%	1999
4	Sacramento Municipal Utility District	Greenergy[b]	8.5%	1997
5	City of Naperville, IL[d]	Renewable Energy Program	8.4%	2005
6	Silicon Valley Power[a]	Santa Clara Green Power[b]	8.1%	2004
7	Pacific Power - Oregon Only[a]	Blue Sky Block[b], Blue Sky Usage[b], Blue Sky Habitat[b]	6.5%	2002
8	River Falls Municipal Utilitiese	Renewable Energy Program[b]	5.8%	2001
9	Stoughton Utilities[e]	Renewable Energy Program[b]	5.2%	2002
10	Lake Mills Light & Water[e]	Renewable Energy Program[b]	5.1%	2002
10	Pacific County PUD	Green Power Tomorrow	5.1%	2002

[a] Marketed in partnership with 3Degrees Group Inc.
[b] Product is Green-e Energy certified.
[c] Marketed in partnership with Green Mountain Energy Company.
[d] Marketed in partnership with Community Energy, Inc.
[e] Power supplied by WPPI Energy.

Table E-4. Green Power Sales as a Percentage of Total Retail Electricity Sales (as of December 2009) (kWh)

Rank	Utility	Program(s)	% of Load
1	Waterloo Utilities[a]	Renewable Energy Program[b]	21.4%
2	Edmond Electric[c]	Pure and Simple	8.1%
3	Portland General Electric[d]	Clean Wind, Green Source, Renewable Future	7.9%
4	City of Palo Alto Utilities[e]	Palo Alto Green[b]	6.9%
5	Austin Energy	Green Choice	6.4%
6	River Falls Municipal Utilities[a]	Renewable Energy Program[b]	6.2%
7	Madison Gas and Electric	Green Power Tomorrow	4.9%
8	Sacramento Municipal Utility District	Greenergy[b]	3.6%
9	Park Electric Cooperative[f]	Green Power Program	3.4%
10	PacifiCorp (Oregon only)[be]	Blue Sky Blockb, Blue Sky Usageb, Blue Sky Habitat[b]	2.8%

[a] Power supplied by WPPI Energy.
[b] Product is Green-e Energy certified.
[c] Power supplied by Oklahoma Municipal Power Authority.
[d] Marketed in partnership with Green Mountain Energy Company.
[e] Marketed in partnership with 3Degrees Group Inc.
[f] Power supplied by Basin Electric Power Cooperative.

Table E-5. Price Premium Charged for New, Customer-driven Renewable Powera (as of December 2009)

Rank	Utility	Resources	Premium (¢/kWh) -
1	Edmond Electric[bc]	Used Wind	0.17
2	OG&E Companyb[d]	Wind	0.28
3	Avista Utilities	Wind, Landfill Gas, Hydro	0.33
4	Park Electric Cooperative	Wind	0.39
5	Arizona Public Service Company[f]	Wind, Geothermal, Biomass, Landfill Gas, Solar	0.40
6	Indianapolis Power & Light Company	Wind	0.42
7	Flathead Electric Cooperative[e]	Wind	0.50
7	Sacramento Municipal Utility Distnct[f]	Wind, Hydro, Biomass, Solar	0.50
9	Xcel Energy (New Mexico)[bf]	Wind, Solar	0.75
10	Emerald People's Utility District	Landfill Gas, Wind, Biomass	0.80

[a] Includes only programs that have installed or announced firm plans to install or purchase power from 100% new renewable resources.

[b] Premium is variable; customers in these programs are exempt or otherwise protected from changes in utility fuel charges.

[c] Power supplied by Oklahoma Municipal Power Authority.

[d] OG&E Company offers two rate structures for its Wind Power program; the lowest premium is for the rate which exempts customers from the fuel charge.

[e] Power is supplied by Basin Electric Power Cooperative.

[f] Product is Green-e Energy certified.

ACKNOWLEDGMENTS

This work was funded by the US Department of Energy's (DOE's) Office of Energy Efficiency and Renewable Energy (EERE). The authors wish to thank Linda Silverman and the EERE technology programs for their support of this work. The authors also wish to thank Blaine Collison of the US Environmental Protection Agency; Steve Dunn of DOE; Bridgett Neely of Green Mountain Energy; Rachael Terada and Alex Pennock of the Center for Resource Solutions; and Lynn Billman, Jim Newcomb, Robin Newmark, and Claire Kreycik of NREL for their thoughtful review of the document as well as Mary Lukkonen of NREL for her editorial support. Finally, the authors thank the many green power marketers and utility contacts who provided the information summarized in this report. Additional information on green power market trends and activities can be found on DOE's Green Power Network Web site at http://greenpower.energy.gov.

REFERENCES

Barbose, G. (8 October 2010). E-mail. Lawrence Berkeley National Laboratory, Berkeley, CA.

Bird, L.; Dagher, L.; Swezey, B. (December 2007). *Green Power Marketing in the United States: A Status Report (Tenth Edition).* NREL/TP-670-42502. Golden, CO: National

Renewable Energy Laboratory. http://apps3.eere. energy.gov/greenpower/resources/pdfs/42502.pdf. Accessed September 2010.

Bird, L.; Hurlbut, D.; Donohoo, P.; Cory, K.; Kreycik, C. (March 2009). *An Examination of the Regional Supply and Demand Balance for Renewable Electricity in the United States through 2015.* NREL/TP-6A2-45041. Golden, CO: National Renewable Energy Laboratory. http://www.nrel. gov/docs/fy09osti/45041.pdf. Accessed September 2010.

Bird, L.; Kreycik, C.; Friedman, B. (October 2008). *Green Power Marketing in the United States: A Status Report (Eleventh Edition).* NREL/TP-6A2-44094. Golden, CO: National Renewable Energy Laboratory. http://www. nrel.gov/docs/fy09osti/44094.pdf. Accessed September 2010.

Bird, L.; Swezey, B. (November 2006). *Green Power Marketing in the United States: A Status Report (Ninth Edition).* NREL/TP-620-40904. Golden, CO: National Renewable Energy Laboratory. http://www.eere.energy.gov/ greenpower/resources/pdfs/40904.pdf. Accessed September 2010.

Bird, L.; Swezey, B. (2005). *Green Power Marketing in the United States: A Status Report (Eighth Edition).* NREL/TP-620-38994. Golden, CO: National Renewable Energy Laboratory.http://www.eere.energy.gov/ greenpower/resources/pdfs/38994.pdf. Accessed September 2010.

Bird, L.; Swezey, B. (September 2004). *Green Power Marketing in the United States: A Status Report (Seventh Edition).* NREL/TP-620-36823. Golden, CO: National Renewable Energy Laboratory. http://www.eere.energy.gov/ greenpower/pdfs/36823.pdf. Accessed September 2010.

Bird, L.; Swezey, B. (October 2003). *Green Power Marketing in the United States: A Status Report (6th Edition).* NREL/TP-620-35119. Golden, CO: National Renewable Energy Laboratory. http://www.eere.energy.gov/greenpower/resources/pdfs/35119.pdf. Accessed September 2010.

Bird, L.; Swezey, B.; Cory, K. (August 2008). *Renewable Energy Price-Stability Benefits in Utility Green Power Programs.* NREL/TP-670-43532. Golden, CO: National Renewable Energy Laboratory. http://apps3. eere.energy.gov/greenpower/resources/pdfs/43532.pdf. Accessed September 2010.

CRS (December 2008). *National Standard Version 1.6.* San Francisco, CA. http://www. greene.org/docs/energy/Appendix%20D_Green-e%20Energy %20National%20 Standard. pdf. Accessed September 2010.

Database of State Incentives for Renewables & Efficiency (DSIRE) (2010). www.dsireusa. org. Accessed August 30, 2010.

Electric Reliability Council of Texas, Inc. (ERCOT) (2009). *ERCOT's 2009 Annual Report On The Texas Renewable Energy Credit Trading Program.* PROJECT NO. 27706. Texas Public Utilities Commission. https://www. texasrenewables.com/staticReports/Annual %20Report/2009 Report.doc. Accessed September 2010.

Emissions & Generation Resource Integrated Database (eGRID) (2010). http://www.epa.gov/ cleanenergy/energy-resources/egrid/index.html. Accessed September 23, 2010.

Evolution Markets (2007)._www.evomarkets.com Accessed December 2007.

Foran, P. (6 July 2010). Personal Communication. The Climate Registry, Los Angeles, CA. Gower, B. (23 August 2010). Personal communication. APX, Inc., San Jose, CA, .

Holt, E.A.; Wiser, R.H. (April 2007). *The Treatment of Renewable Energy Certificates, Emissions Allowances, and Green Power Programs in State Renewables Portfolio*

Standards. LBNL-62574. Lawrence Berkeley National Laboratory, Berkeley, CA. http://eetd.lbl.gov/ea/ems/ reports/ 62574.pdf. Accessed September 2010.

PJM-GATS (2010). Solar Weighted Average Price. https://gats.pjmeis.com /myModule/rpt /myrpt.asp?r=230. Accessed September 23, 2010.

Power to Choose (2010). www.powertochoose.org. Accessed September 23, 2010.

Schuyler, K. (7 July 2010). Personal communication. PJM Environmental Information Services, Inc., Norristown, PA.

Spectron Group (2010). Data provided by Christian Thalacker. London, UK. www.spectron group.com

SRECTrade (2010). http://www.srectrade.com/about.php. Accessed September 23, 2010.

Swezey, B.; Bird, L. (August 2000). *Green Power Marketing in the United States: A Status Report (5th Edition)*. NREL/TP-620-28738. Golden, CO: National Renewable Energy Laboratory. http://www.eere.energy.gov greenpower/resources/pdfs/28738.pdf. Accessed September 2010.

Swezey, B.; Bird, L. (August 1999). *Information Brief on Green Power Marketing (4th Edition)*. NREL/TP-620-26901. Golden: CO: National Renewable Energy Laboratory. http://www.eere.energy.gov/greenpower /resources/pdfs/26901.pdf. Accessed September 2010.

Terada, R. (9 October 2010). Personal communication. Center for Resource Solutions, San Francisco, CA.

World Resources Institute (WRI) and World Business Council for Sustainable Development (WBCSD) (2004). The Greenhouse Gas Protocol: A Corporate Accounting and Reporting Standard (Revised Edition). http://www.wri.org/publication/greenhouse-gas-protocol-corporateaccounting-and-reporting-standard-revised-edition.Accessed September 2010.

End Notes

[1] The term "green power" generally refers to electricity supplied in whole or in part from renewable energy sources, such as wind and solar power, geothermal, hydropower (typically low-impact or small hydro), and various forms of biomass.

[2] Under these programs, consumers can buy renewable energy from independent renewable energy marketing companies without switching their electricity service from the default or standard-offer service provider.

[3] States with competitive offerings include Connecticut, Delaware, Illinois, Maine, Maryland, Massachusetts, Michigan, New Jersey, New York, Pennsylvania, Rhode Island, Texas, and Washington, D.C.

[4] Green power market data for previous years are available in Bird et al. (2009), Bird et al. (2008), Bird et al. (2007), Bird and Swezey (2006).

[5] US electricity sales totaled 3,732 billion kWh in 2008 (2009 data are not yet available), according to the US Energy Information Administration (EIA). See http://www.eia.doe. gov/cneaf/electricity/epa/epat7p2.html. The remaining renewable energy generation is rate-based by utilities or used to meet renewable portfolio standards.

[6] Other sources include REC certifiers, REC tracking systems (see ERCOT 2009), and press releases describing large voluntary green power purchases.

[7] Estimates of the above market value of green power sales are determined by multiplying green power sales in kWh in three subsectors (utility green pricing programs, residential competitive markets, and nonresidential competitive and REC market) by a low and high estimate of prices in each of the sectors.

[8] The REC sales figures reflect sales to end-use customers separate from electricity. RECs bundled with electricity and sold to end-use customers through utility green pricing programs or in competitive electricity markets are counted in other categories.

[9] Capacity estimates are calculated based on reported green power kilowatt-hours sales assuming capacity factors for each renewable resource type. For wind, a capacity factor of 33% was assumed, 90% for landfill gas, 80% for biomass, 96% for geothermal, 40% for hydroelectric, and 15% for solar electric.

[10] "New" renewable energy capacity defined here is capacity that was sourced from renewable energy systems that were built or repowered after January 1, 1997.

[11] It is important to note that there is greater uncertainty in our customer estimates for competitive and REC markets because of data limitations. For more detailed estimates by state for 2007 and 2008, see data from US EIA 2009 in Appendix C. Generally, our estimates are consistent with the EIA estimates when adjusted for customers in Ohio, who participated in community aggregations in 2005 and earlier. We excluded these customers from our estimates because they purchase products with very low renewable energy content (1%–2%).

[12] Arizona and Wisconsin are the only states that explicitly allow renewable energy purchased through voluntary programs to also count toward the RPS. (Holt and Wiser 2007) However, no utilities in these states that have reported data on voluntary programs to NREL have elected to count green power sales toward RPS compliance.

[13] For additional details on the treatment of voluntary green power purchases in state RPS policies, see Holt and Wiser (2007).

[14] Although RPS policies generally allow pre-existing renewable energy generation sources (i.e., those installed prior to the adoption of the RPS) to meet their targets, the estimates presented here reflect only the amount of new renewable energy generation that these policies are expected to stimulate. These figures are compared to the voluntary market estimates because voluntary markets primarily support generation from new renewable energy projects (i.e., those installed after voluntary green power markets were established). Estimates of compliance market demand assume that RPS targets are fully met.

[15] These states include Colorado, Delaware, Iowa, Maine, Montana, New Mexico, Oregon, Virginia, and Washington (DSIRE 2010). Maine passed legislation in 2009 that requires the Public Utilities Commission (PUC) to develop a program offering green power; the PUC is still in the process of developing the program.

[16] For example, a small number of utilities exempt green pricing customers from monthly or periodic fuel charges imposed to pay higher than expected fossil-fuel costs. For a more detailed discussion of this topic, see Bird et al. (2008).

[17] NREL attempted to contact all utility green pricing programs and received data for about 60% of programs in 2009, including all of the major programs. The remaining programs, which are smaller in size, do not have a large impact on overall participant numbers. Wherever possible, other sources and previously reported data were used to estimate data gaps.

[18] NREL issues five different Top 10 lists based on total sales of renewable energy to program participants, total number of customer participants, customer participation rates, green power sales as a fraction of total utility sales, and the premium charged to support new renewable energy development. These lists can be found in Appendix E or at http://apps3. eere.energy.gov/greenpower/markets/pricing.shtml?page=3.

[19] The Florida Public Service Commission initially acted to discontinue the program as a result of concerns over the amount of program revenues spent on marketing compared to expenditures on the renewable energy resources used to supply the program, as well as its support for out-of-state resources. However, the final basis for the decision to terminate the program, after a subsequent program audit, was related to the commission's assessment that a voluntary program was not needed after the Florida Legislature mandated an RPS. By Order No. PSC-08- 0600-PAA-EI, issued September 16, 2008, in Docket No. 070626-EI, the commission terminated the program. http://www.floridapsc.com/library/filings/08/ 087 20-08/08-0600.ord.doc.

[20] Capacity estimates in 2008 and 2009 are calculated based on reported green power kilowatt-hours sales assuming capacity factors for each renewable resource type. For wind, a capacity factor of 33% was assumed, 90% for landfill gas, 80% for biomass, 96% for geothermal, 40% for hydroelectric, and 15% for solar electric. Estimates of megawatts prior to 2008 were higher on a relative basis due to the capacity factor assumed for wind. In prior years, a 30% capacity factor was assumed, but in 2008, estimates of megawatts were based on a 33% capacity factor to reflect improvements in capacity factors as a result of the movement toward larger turbines as well as greater reliance on projects in areas with strong wind resources. For every 1 million MWh, this accounts for a discrepancy of 35 MW of capacity in the estimates.

[21] For an up-to-date list of products offered by competitive green power marketers, see the US Department of Energy's Green Power Network Web site at: http://apps3.eere.energy. gov/ greenpower/markets/ marketing. shtml?page=1.

[22] We do not include Oregon and Virginia in this list. In Oregon, only large commercial and industrial customers are able to switch to competitive green power providers; residential and small commercial customers have access

to green power options offered by the incumbent utilities, which we categorize as green pricing. In Virginia, at least one retail electricity provider provided green power options in 2007 and earlier but does not do so currently.

[23] For an up-to-date list of companies offering REC-based green power products, see the US Department of Energy's Green Power Network Web site at: http://apps3.eere.energy.gov/ greenpower/markets/certificates.shtml?page=1. For a list of REC suppliers serving commercial or wholesale customers, see: http://apps3.eere.energy.gov/greenpower/ markets/ certificates.shtml?page=4.

[24] Administered by the San Francisco-based Center for Resource Solutions, the Green-e Energy program certifies retail and wholesale green power products that meet its environmental, product content, and marketing standards. For details on the Green-e Energy National Standard, see the Green-e Web site at: http://www.greene.org/.

[25] See the EPA's Green Power Web site at: http://www.epa.gov/greenpower.

[26] The EIA figures include customers in both utility green pricing programs and competitive market programs.

[27] The data is published annually in a report by ERCOT to the Texas PUC. Retirements from the most current year (2009) are reported in aggregate, while retirements from the previous year (2008) are reported by marketer. These voluntary retirements include both bundled and unbundled REC purchases. In order to provide an accurate estimate of competitive market sales in Texas, which we incorporate into total competitive market sales, the 2008 data reported to the Texas PUC were adjusted to account for marketers and utilities that had already provided data to NREL. Of this leftover total, NREL included sales of bundled RECs into the competitive market category. For 2009, data are not yet available by marketer; in order to provide a conservative estimate of the competitive marketer, the same amount of sales added in 2008 were also added to 2009 figures. 2009 data may need to be modified if individual marketer data for 2009, due to be released in May 2011, are different from our current estimate.

[28] For example, the EPA Green Power Partnership reports that the majority of its Top 25 partners purchases RECs (see Appendix A). For more information, see http://www.epa. gov/greenpower/. In addition, the Green Power Market Development Group promotes the purchase of RECs among its members. For more information, see the organization's Web site at: http://www.thegreenpowergroup.org/.

[29] Capacity estimates are calculated based on reported green power kilowatt-hours sales assuming capacity factors for each renewable resource type. For wind, a capacity factor of 33% was assumed, 90% for landfill gas, 80% for biomass, 96% for geothermal, 40% for hydroelectric, and 15% for solar electric.

[30] "New" renewable energy capacity defined here is capacity that was sourced from renewable energy systems that were built or repowered after January 1, 1997.

[31] The Green Power Network tracks GHG offset providers and products that are available nationally and derived, at least in part, from US-based renewable energy generation projects.

[32] In February 2008, the Center for Resource Solutions certified its first retail offset products under Green-e® Climate, a consumer-protection program requiring verification of GHG reductions based on product-level certification that ensures that emissions reductions come from projects verified and certified under project standards that meet a high standard of quality, that the emissions reductions are not being double-sold, and that consumers are being given full and accurate information. Sellers must undergo a yearly audit to ensure their supply of offsets matches their sales, and a twice annual review of website and marketing materials to ensure compliance with Green-e Climate's consumer disclosure and truth in advertising requirements. The Green-e Climate Protocol for Renewable Energy is a project standard that establishes the eligibility requirements for renewable energy projects in the United States that wish to supply Green-e Climate certified offsets, including methodologies used to assess additionality and calculate the emission reductions, and other requirements related to tracking, prevention of double counting and double claiming, and verification. The Protocol requires that the RECs associated with the renewable energy generation producing the emissions reductions that are certified under Green-e Climate be retired as part of the substantiation for any carbon offset claim and not resold in the voluntary green power markets or used for compliance with renewable energy standards. The seller must verify that the attributes are only sold once and not double counted. For more information, see the Protocol at http://www.green-e.org/docs/climate/Green-eClimate ProtocolforRenewableEnergy.pdf.

[33] The EPA's national average electricity emissions factor for non-baseload generation (eGRID 2010) was used to estimate the equivalent in MWh for companies that did not report their sales in MWh.

[34] The Environmental Resource Trust/Winrock International verifies carbon offsets in partnership with the American Carbon Registry. The American Carbon Registry allows flexibility for members to choose among methodologies set out by the Clean Development Mechanism and the Voluntary Carbon Standard. A carbon offset is considered an emissions reduction ton (ERT) if it is real, additional, permanent, and that ownership is

incontestable. After verification, the Registry assigns each offset a unique serial number. For more information on the ERT certification, see http://www.winrock.org/common/files/Solution Stories/acr capabilities.pdf.

[35] The CCX guidelines for carbon offsets sourced from renewable energy generation were established in 2006. To qualify, renewable energy systems must have been activated on or after January 1, 2005. Project proponents must demonstrate ownership rights associated with the environmental attributes (i.e., they must not have sold the RECs or used them for compliance purposes). Under the verification process, for CCX offsets to be issued, the RECs are surrendered to and retired by CCX. For more information on the CCX guidelines, see http://www.chicagoclimatex.com/news/publications/pdf/CCXRenewableOffsets.pdf

[36] The data can be queried online at: https://gats.pjm-eis.com/myModule/rpt/myrpt.asp?r=230.

[37] http://www.srectrade.com/.

[38] http://www.flettexchange.com/.

[39] http://www.pjmenvirotrade.com.

[40] PJM Interconnection covers all or parts of Delaware, Illinois, Indiana, Kentucky, Maryland, Michigan, New Jersey, North Carolina, Ohio, Pennsylvania, Tennessee, Virginia, West Virginia, and Washington, D.C.

[41] www.njcep.com/srec.

[42] http://paaeps.com/credit/pricing.do.

[43] ERCOT's Annual Report on the Texas Renewable Energy Credit Trading Program is available for download here: https://www.texasrenewables.com/staticReports/ Annual%20 Report/2009Report.doc.

[44] For further information, see http://www.epa.gov/climateleaders/index.html. Accessed July 6, 2010.

[45] For additional information on Climate Leaders protocols for offsets and green power, see http://www.epa.gov/climateleaders/resources/optional-module.html. Accessed July 6, 2010.

[46] For further information on Climate Leaders guidance on GHG accounting for green power and RECs, see http://www.epa.gov/climateleaders/documents/greenpowerguidance.pdf.

[47] For further information, see http://www.theclimateregistry.org.

[48] For further information, see page 101. http://www.theclimateregistry.org/downloads/GRP.pdf.

[49] For more information on the electric sector protocol, see http://www.theclimateregistry.org/ resources/protocols/ electric-power-sector-protocol/. Accessed July 6, 2010.

[50] For further information, see The Local Government Operations Protocol:http://www. icleiusa.org/programs/ climate/ghg-protocol/local-government-operations-protocol. Accessed August 9, 2010.

[51] It should be noted that currently, voluntary purchases have minimal impact on overall emissions rates.

[52] For further information, see "Draft Guidance for Federal Greenhouse Gas Accounting and Reporting," http://www.whitehouse.gov/administration/eop/ceq/sustainability/fed-ghg. Accessed August 9, 2010.

In: The Consumer Component in Renewable Energy Use
Editors: Georgia M. Lee and Russ L. Cooper
ISBN 978-1-62081-425-3

Chapter 3

COMMUNITY RESPONSE TO CONCENTRATING SOLAR POWER IN THE SAN LUIS VALLEY: OCTOBER 9, 2008 – MARCH 31, 2010*

B. C. Farhar, L. M. Hunter, T. M. Kirkland and K. J. Tierney

PREFACE

The study "Social, Economic, and Policy Factors Affecting the Siting of a 100-MW Concentrating Solar Power (CSP) Facility in Colorado's San Luis Valley" was a joint effort of the University of Colorado (CU) and the National Renewable Energy Laboratory (NREL). The CU study, conducted at the Institute of Behavioral Science, was funded by NREL and the Energy Initiative—now the Renewable and Sustainable Energy Institute (a joint institute of CU and NREL).

The period of performance for the study was originally November 2006 through February 2009. The NREL portion of the study was funded by contract to co-principal investigator Barbara Farhar, the first author of this report. The period of performance for the contract was October 9, 2008 through March 31, 2010.

A Research Advisory Committee provided guidance on key questions and groups to include in the study. Committee members were: Mark Mehos, NREL; Fred Morse, Chairman, Utility-Scale Solar Power Division, Solar Energy Industries Association; Mike Wisdom, San Luis Valley Regional Development Group; Mike Bowman, 25 x '25; and Marty Smith, Xcel Energy.

Part of this report is based on interviews with representatives of 25 key stakeholder groups within and outside the San Luis Valley. The CU principal investigator was Kathleen J. Tierney, Ph.D., and Lori M. Hunter, Ph.D., was also a co-principal investigator. Tracy M. Kirkland, a doctoral candidate in the Department of Sociology, provided graduate research

---

* This is an edited, reformatted and augmented version of a National Renewable Energy Laboratory Subcontract Report NREL/SR-550-48041 publication, dated June 2010.

assistance on the joint project. Other research assistance was provided to Dr. Farhar by Gregg Eisenberg (Iron Mountain Consulting) who did the primary work on the transmission issues after the interviewing ended, and Linda Giudice (graduate student, University of Colorado at Boulder) who did the primary work on the Bureau of Land Management Programmatic Environmental Impact Statement process. Therese Edwards and Teresa Foster provided research assistance on coding. Sarah Leshan provided editorial support. Many helpful suggestions on the manuscript were made by reviewers Adam Reed, Center for Energy and Environmental Security, University of Colorado Law School; Timothy Schoechle, Consultant; Fred Morse; and Bill Mansheim, Adams State College. Any errors are the responsibility of the authors.

> "The Valley is an economically depressed region and CSP represents the biggest economic development opportunity that has happened in the world. There is a juxtaposition of the vulnerability of the Valley because of its poverty and its unique cultural heritage on the one hand and this great opportunity on the other."
>
> - A stakeholder from outside the Valley

> "I look over this Valley over the last 20 years...when we first got here you could see Alamosa, you could see Monte Vista...you could see the population centers. Now, it's just all lights at night. No one seems to want to turn anything off...it's sad. With this new transmission line that is coming in, that's another indicator that there are just going to be more lights. Do we need more light? Are we so afraid of the dark as a culture that we have to have lights on 24/7? I think we are. We're a bunch of fraidy cats...this transmission marketing scheme is all fear-based."
>
> - A stakeholder from within the Valley

> "Our people are very vocal...if they came in and worked with us, there wouldn't be much opposition. But if an outside group tried to come in and strong-arm us, this is a tough community. We want to protect the environment and to protect against big government and corporations. If it is done properly and through the right channels...so, really, it is how it is done."
>
> - A stakeholder from within the Valley

EXECUTIVE SUMMARY

This report is about the social acceptance of utility-scale renewable energy projects. In particular, it focuses on concentrating solar power (CSP) plants in the San Luis Valley (SLV or Valley), a large, sunny, high desert in south central Colorado, located within 200 miles of greater Denver.

The research focused on social factors that may facilitate and impede the adoption and implementation of CSP—a technology that captures the sun's thermal energy using curved mirrors to focus sunlight onto a high temperature receiver.

During the winter of 2008-2009, qualitative in-depth interviews were conducted with a purposive sample of 25 CSP-related stakeholders inside and outside the SLV of Colorado. Interviews focused on the perceived advantages and disadvantages of siting a hypothetical 100-MW CSP facility in the Valley, the level of community support and opposition to CSP development, and related issues, such as transmission.

Key groups included education, advocacy, and professional organizations within and beyond the Valley; SLV economic groups; government and regulatory agencies, organizations, and representatives; farmer and rancher groups; CSP industry groups; environmental organizations; and utilities. Content analysis resulted in five categories of perceived advantages and disadvantages: economic, environmental, technological, social-psychological, and regulatory.Perceived benefits included potential improvements in the SLV economy, job creation, local tax property and other tax revenues to local counties, provision of clean energy, putting land to best use, reducing greenhouse gas emissions, addressing climate change, improving the security and diversity of power supplies, reducing dependence on imported energy, providing dispatchable energy, a source of community pride, positive effects on community identity, and helping to meet Colorado's renewable portfolio standard.

If the not-in-my-backyard (NIMBY) syndrome is defined as a community's hypocritical espousal of renewable energy while rejecting its geographical presence, then no evidence was found for a “NIMBY-syndrome” response to CSP. Concerns included the need for water for CSP development, land-use intensity, concern that a boom-bust cycle might result from CSP plant construction and that permanent jobs would be minimal, concern that transmission would be needed to export CSP power, hesitation about the efficacy of CSP technology, intermittency of solar power, concern that CSP is difficult for residents to conceptualize, and concern that it is difficult for Valley residents to deal with Xcel Energy. Valley stakeholders emphasized concerns somewhat differently than did those outside the Valley—Valley stakeholders emphasized environmental and social-psychological benefits more than stakeholders outside the Valley, whereas outside-the-Valley stakeholders tended to emphasize technological benefits (especially energy storage). Both groups emphasized the economic advantages for the SLV, but Valley stakeholders focused more than outsiders on boom-bust cycle concerns.

Stakeholders indentified supporters and opponents of CSP and transmission development, including 125 different stakeholder groups representing agriculturists, farmers, ranchers, and their associations; CSP developers and their trade associations; environmental organizations; SLV economic development organizations; educational institutions; elected officials, federal electricity agencies; federal government entities; state government entities; local and regional government entities and non-government organizations; landowners; law firms; private companies; the public; solar and sustainable resource advocates; spiritual groups; utilities and transmission line owners; and water interests.

Three different models of CSP facility and transmission siting decision-making were elicited from stakeholder interviews. These were (1) SLV Power Authority decision-making, (2) county permitting, and (3) permitting of utility-scale solar projects on public lands. Some Valley stakeholders supported local development and management of distributed energy systems.

In May 2009, just after the interviewing was completed, Xcel Energy and Tri-State Electric Generation and Transmission Association filed for a Certificate of Public Convenience and Necessity (CPCN) for a 230-kV transmission line from Calumet Substation into the San Luis Valley at Highway 160 over La Veta Pass. A controversy erupted over the proposed San Luis Valley-Calumet-Comanche transmission line; an adversarial proceeding ensued as part of the Public Utilities Commission decision process on the utilities' application. Leading the opposition to the line was the Trinchera Ranch.

The Bureau of Land Management, which has a backlog of applications for permits to site large-scale solar facilities on public lands, also conducted public hearings for a Programmatic Environmental Impact Statement (PEIS) for solar development on public lands in six western states, including the SLV. Decisions on all matters were pending as of March 31, 2010, when this report was completed.

Widespread support was found for "solar energy" development in the SLV and Valley stakeholders took a great deal of pride in the SunEdison 8.22-MW PV system near Mosca. However, the Valley's residents lack knowledge on renewable energy technologies; for example, widespread confusion was found among stakeholders on the difference between PV and solar thermal electric systems. Education on renewable energy is needed for the Valley's stakeholders and residents.

Siting of CSP facilities and of the transmission line involve complex decision processes with a confusing mix of entities and procedures.

Public meetings and hearings have been held by independent power producers, the Bureau of Land Management, the Rural Utility Service, the Colorado Public Utility Commission, and others.

Supporters of the transmission line and opponents of it have testified at these meetings. Some Valley stakeholders feel overwhelmed by the attention and demands on their time made by developers and decision makers.

An integrated and transparent decision process is needed that fully and effectively integrates federal and state policies with concerns of the local area.

State policy recommendations based on the findings include developing education programs for SLV residents, integrating Valley decision makers into an energy-water-land group, providing training for Valley decision makers, offering workforce training, evaluating models of taxation, and forming landholder energy associations. In addition, the SLV could become a laboratory for new approaches to CSP facility and transmission siting decision-making.

Those outside the Valley must address community concerns and engage Valley residents in CSP decisions.

The opportunity to protect vulnerable communities and to engage them in CSP and transmission decisions, benefitting from their expertise rather than treating them as adversaries, should parallel in significance the investment in solar technology.

Abbreviations and Acronyms

ARRA	American Reinvestment and Recovery Act
BLM	Bureau of Land Management
CEDA	Clean Energy Development Authority
CFR	Code of Federal RegulationsCOSEIA Colorado Solar Energy Industries Association
CPCN	Certificate of Public Convenience and Necessity

CRES	Colorado Renewable Energy Society
CSP	concentrating solar power
CU	University of Colorado
DG	distributed generation
DOE	US Department of Energy
DOI	US Department of the Interior
DSM	demand-side management
DWP	Los Angeles Department of Water and Power
EMFs	electro-magnetic fields
ERZ	Energy Resource Zone
FERC	Federal Energy Regulatory Commission
FLPMA	Federal Land and Policy and Management Act
GEO	Colorado Governor's Energy Office
IOUs	investor-owned utilities
IPP I	ndependent power producer
IPPs	independent power producers
ITC	Investment Tax Credit
kWh/m	kilowatt-hours/meter
LGIT	large generator integration process LULUs “locally unwanted land uses”
MOU	Memorandum of Understanding MW megawatt
NEPA	National Environmental Policy Act
NERC	North American Electric Reliability Corporation
NIMBY	“not in my backyard”
NOI	Notice of Intent
NREL	National Renewable Energy Laboratory
PV	photovoltaic
PEIS	Programmatic Environmental Impact Statement
PPA	power purchase agreement
PUC	Colorado Public Utilities Commission
REC	Rural Electric Cooperative
RFR	right of first refusal
RMPs	Resource Management Plans
ROW	rights-of-way
RPS	renewable portfolio standards
RUS	Rural Utility Services
SESAs	Solar Energy Study Areas
SLB	State Land Board
SLV	San Luis Valley
SWEEP	Southwest Energy Efficiency Project USDA US Department of Agriculture
USFS	US Forest Service
WAPA	Western Area Power Administration WECC Western Electric Coordinating Council
WGA	Western Governors' Association
WRA	Western Resource Advocates

1. Introduction and Background

1.1. Introduction

This report is about the social acceptance of utility-scale renewable energy projects. In particular, it focuses on concentrating solar power (CSP) plants in the San Luis Valley—a large, sunny, high desert in south central Colorado, located within 200 miles of greater Denver. Although this research is primarily focused on generation, transmission issues emerged as an important aspect because CSP is most feasible with a means to export the power from the Valley. Although the transmission issue that emerged related to CSP, it ultimately pertains to any form of utility-scale generation.

Social acceptance of utility-scale renewable energy projects has long been recognized as one of the most powerful factors in achieving renewable energy goals.[1] However, barriers to the social acceptance of renewable energy facilities present new challenges that must be addressed if US renewable energy objectives are to be met. Unfortunately, little systematic knowledge exists on public receptivity to the siting of different types of renewable energy facilities, even though facilities differ in characteristics likely to influence public acceptance, such as their appearance and the size of their geographic footprints. For example, recent research has focused on the siting of large wind turbines,[2] but it is not clear how findings from research on that particular technology might be transferable to other utility-scale renewable energy facilities.

The research in this report analyzes social factors that may facilitate and impede the adoption and implementation of another type of renewable energy technology—concentrating solar power (CSP). CSP technologies capture the sun's thermal energy using curved mirrors to focus sunlight onto a high temperature receiver.

The first commercial CSP plants have been in operation in California since the mid-1980s, providing hundreds of megawatts (MW) of the world's lowest-cost solar power. The many types of systems under development (parabolic troughs, power towers, dish engine systems, and linear Fresnel systems) vary in terms of energy concentration technologies used—such as concentration devices, conversion methods, and storage options— and in terms of their aesthetic features.

Social factors are important in siting decision-making. Although the conventional energy industry is well acquainted with prevailing energy facility siting decision-making procedures, the renewable energy community is still climbing the learning curve. We found that a major obstacle to CSP and transmission siting is the complexity of decision-making on siting central CSP facilities and new transmission lines. Decision roles appear to be compartmentalized by various groups; the decision processes for this new technology—which are not well understood by stakeholders and the public—are not even fully understood by decision makers themselves. An integrative decision role seems lacking.

Furthermore, contemporary decision models themselves evolved in an earlier era of conventional energy development and may be peculiarly suited to the specific needs of those industries. Therefore, new decision models are needed that are consonant with the needs of CSP siting. For example, competition for extensive real estate is a key issue for CSP in the San Luis Valley (SLV).

New transmission lines are needed in areas where they have never been located to bring CSP electricity to demand centers. This issue can be controversial, even in communities that espouse renewable energy development.

Key questions about how the SLV can benefit from CSP development are identified in this report and remain to be answered. Transmission siting also ties into the tension in communities between preferences for distributed photovoltaic (PV) systems, medium-scale PV systems, and utility-scale centralized solar energy systems (PV or CSP).

This report documents experiences and perspectives of stakeholders within and outside the Valley.

Renewable energy and transmission stalwarts are working diligently to overcome institutional barriers to renewable energy development and transform existing siting decision processes. In this, they are challenging conventional energy industries and working to forge new transmission lines on lands where none existed before to bring far-flung wind, solar, geothermal, and other renewable resources to energy demand centers. Inevitably, conflicts arise and adversarial proceedings ensue.

The National Environmental Policy Act (NEPA) requires analysis, review, and mitigation of negative socio-economic impacts of major proposed actions on federal lands that include federal funding and public participation in decision-making.

Some of the renewable energy community—accustomed to thinking of itself as offering energy solutions superior from an environmental perspective to carbon-based electricity—may believe that their proposed projects should be exempted from major environmental and socio-economic review processes, especially in light of what is increasingly identified as a global climate emergency.

From this standpoint, any local community identified by scientists, policy makers, and industry as highly suitable for solar or wind development should welcome, or at least not resist, proposed renewable energy facilities in their areas that have the end goal of decarbonizing the electricity supply.

Nevertheless, local communities are embedded in familiar landscapes, dependent on local economic interests, and reliant on a prevailing cultural heritage on which proposed projects may uncomfortably impinge. Moreover, they raise questions about local control and about what a project can bring to them economically—particularly as projects are carried out by private corporations for profit, some of which can be expected to leave the local area. They ask what the proposed project will mean to the identity of the local community.

These broader issues of utility-scale renewable energy project development apply to CSP specifically and to facility and transmission siting in the SLV.Although federal action has been taken to speed solar development on public lands, the laws and procedures governing socio-economic and environmental review and mitigation of adverse socio-economic impacts still need to be recognized and respected within and beyond the federal lands.

1.2. Sociological Aspects of Renewable Energy Siting

Research on opposition to facility siting in various contexts has been reported in the social science literature, in which researchers present data and arguments about the significance and meaning of such terms as the NIMBY ("not in my backyard") and LULUs ("locally unwanted land uses") syndromes. Research on public participation has long suggested that public—and especially stakeholder—participation in the decisions surrounding proposed facility siting will increase the public's satisfaction with the decision process and its ultimate acceptance of the facility.

The social acceptance of technology has been identified as having three dimensions—sociopolitical acceptance, community acceptance, and market acceptance. The research reported here focuses primarily on community acceptance, which refers to "specific acceptance of siting decisions and renewable energy projects by local stakeholders, particularly residents and local authorities."[3] Several factors are thought to be related to community acceptance: (1) distributive justice (the distribution of costs and benefits of a renewable energy facility); (2) procedural justice (the inclusiveness of the decision-making process); and (3) the amount of trust placed in actors from outside the community.[4]

Although social rejection of energy projects such as nuclear power plants and waste repositories is well documented, barriers to social acceptance of "clean" energy facilities present new challenges that need to be addressed if carbon-reduction goals are to be met and if renewable energy policies are to succeed.[5] Little empirical evidence supports the claim that NIMBYism is responsible for a difference between favorable attitudes toward renewable energy technologies and community response to proposed facilities in a local area.[6] Instead, Wolsink argues that the NIMBY label is used to discredit opposition concerns and has "become a great hindrance to the handling of critical attitudes."[7] We find that NIMBY and LULU explanations are oversimplifications of concerns expressed by local stakeholders. Instead, our findings reveal that local concerns are complex, typically incorporating several elements of social, cultural, technological, political, and environmental elements.

We also find that local and outside stakeholders, with their different perspectives and understandings, are involved in complex and sometimes tense interactions that will result in significant legal and regulatory decisions that will in turn permanently affect the texture and quality of everyday life for the residents of the San Luis Valley. These situations also resonate in other potential large-scale solar energy siting locales.

1.3. Background

The background section has two parts. The first part deals with the reason for the study—the National Renewable Energy Laboratory's designation of the SLV as a critical solar resource in the State of Colorado.

The second part of the background section deals with the historical, cultural, and demographic characteristics of the research setting. These characteristics are germane to understanding the complex nature of the Valley's response to projects, why it is a prime solar siting area, and how contemporary concerns are rooted in historical development.

1.3.1. CSP Technology and the SLV

In January 2006, the Western Governors' Association (WGA) Solar Task Force Report identified the two central-station solar technologies under their consideration: (1) solar thermal electric and (2) PV generators. The vast majority of central-station solar projects underway or deployed are solar thermal electric (CSP).

Although the Task Force believed that central-station flat-plate PV deployment will become more commonplace in the future, distributed PV generation has a much higher energy market value than central station PV. Therefore, the Task Force focused on CSP for its discussion of central-station technology and PV for its discussion of distributed energy technology.

According to the report, "The southwestern United States possesses a world-class, well-distributed, and nearly untapped solar energy resource,"[8] a conclusion based on an NREL analysis.[9]

The WGA identified optimal sites for CSP in six western states (Arizona, California, Colorado, New Mexico, Nevada, and Utah). The NREL GIS analysis used the following filters to determine which lands were suitable for CSP plants:

- Only lands with an average daily solar resource of 6.75 kilowatt-hours/meter (kWh/m) or above
- 500 contiguous acres of land minimum (5 acres/MW for 100-MW CSP plants)
- Land with 1 percent or less slope
- Excluded designated urban areas, national parks, national preserves, wilderness areas, wildlife refuges, or water areas.[10]

Given these parameters, NREL identified the SLV as the premier site for CSP facility siting in the state of Colorado.[11] The WGA Task Force estimated the target goal for CSP electricity production at 20% of market penetration for the six Southwestern states. Calculations of CSP deployment by state resulted in an estimated 5,300 GW peak demand growth in Colorado (of 34,100 GW in the six-state region). The CSP allocation of this demand growth in Colorado was 0.1 GW.[12]

The US Department of Energy (DOE)[13] identified seven Western States in the Southwest (Arizona, California, Colorado, New Mexico, Nevada, and Utah) with the combination of solar resource and available suitable land to generate up to 6,800 GW of electricity. (The electric generating capacity of the United States is currently greater than 1,000 GW.) A land area of 2,124 square miles in Colorado's SLV was identified as part of this prime national solar resource.

The DOE report noted that CSP technology operates most efficiently where the skies are clear (as they are in the arid SLV).

1.3.1.1. CSP Benefits

The DOE and WGA reports identified several benefits of CSP development, as follows:[14]

- Adding CSP to the generation mix of a state expands domestic energy supply.
- CSP provides better air quality than other energy sources, avoiding emissions.
- CSP provides a hedge against possible future carbon taxes.
- CSP is flexible enough to be deployed as large central-station power plants or as smaller-scale distributed resources.
- CSP provides local economic development benefits (increases in jobs and electricity price stability) as long-term fuel costs associated with natural gas and coal are replaced by labor costs of operations and maintenance.
- Utilities have operated central station power plants for decades and are more comfortable with them than with distributed systems.

- CSP deployment helps meet state renewable portfolio standards (RPS) requirements.
- CSP adds diversity to the energy supply, increases grid security, increases use of local resources, and cannot deplete the solar resource.

The Task Force noted that CSP can be configured as a "hybrid" power facility with auxiliary gas-fired equipment to achieve full power and remove intermittency from operation. It noted: "The ability to dispatch power during peak demand periods makes CSP an ideal renewable energy technology for the Southwest." [15]

A California economic benefits study[16] found that building between 2,000 and 4,000 MW of CSP would benefit California in the following ways:

- Provide $7 to $13 billion in new investment, of which $2.8 to $5.4 billion would be spent in California
- Increase gross state product by $13 to $24 billion
- Create 1,500 to 3,000 permanent jobs.

Mark Mehos (NREL), in a National Public Radio interview broadcast on March 15, 2009, said that CSP is a "load following" technology allowing utilities to match electricity generation to the load requirements of human activity; for example, energy can be stored overnight to meet morning loads. CSP prices are said to be fixed because there are no fuel components to alter them. In the same interview, Fred Morse (Solar Energy Industries Association) said that the types of jobs created by CSP are broad, including those in engineering, optics, R&D and operations and maintenance. Specific positions would include (but not be limited to) electricians, heavy equipment operators, and mirror cleaners.

1.3.1.2. Policy Recommendations

The WGA Task Force made several key policy recommendations relevant to CSP development. These included the following.

- *Streamlined permitting.* Permitting and siting large power plants is costly and time-consuming.[17] The Task Force called for standardized, streamlined fast-track permitting procedures for CSP plants "while retaining the need to provide for public oversight and protection."[18]
- *Renewable portfolio standards.* Renewable energy portfolio standards are known to have stimulated central station solar developments.[19] These standards require that a specific portion of a state's electricity consumption be met by renewable energy by a certain year (in Colorado, the RPS has been increased to 30% by 2020 [H.B. 1001]).
- *Investment tax credit (ITC).* On October 8, 2008, the Emergency Economic Stabilization Act (P.L. 110-343) extended the solar 30% Federal ITC to utilities through December 31, 2016 and extended it to utilities. This extension allows time to design, permit, finance, and build central solar power plants, and would give a $ 0.03/kWh price reduction for electricity from CSP plants. 20
- *Sales and property tax exemptions.* The WGA report recommended that central solar plants be exempted from sales and property taxes; they said that the loss of these taxes

would be offset by new tax revenues generated by economic activity caused by the plants.[21]

- *Aggregation of plant orders and project bids.* The WGA recommended that state public utility commissions (PUCs) extend power purchase agreements (PPAs) to 30 years.[22] Aggregating plant orders would step up early-stage cost reductions.

DOE[23] stated that it did not necessarily agree with all of the WGA recommendations. DOE announced on September 19, 2009, that it will invest up to $35 million over the next 4 years in 15 CSP facilities, which will result in up to a $67.6 million investment.[24]

Other than to mention that permitting can be costly, the reports reviewed did not analyze the dynamics and concerns of local communities relative to CSP facility siting and transmission siting. They did not review land-use impacts. The reports focused on solar resource, technology, jobs impacts, economic multiplier effects, and environmental effects.

1.3.2. The Research Setting

The San Luis Valley of Colorado is a historically and culturally sensitive area. Located in south-central Colorado and ringed by 14,000-foot-high peaks of the San Juan and Sangre de Cristo mountain ranges, the Valley shares a southern border with New Mexico. The six counties of the Valley (Alamosa, Conejos, Costilla, Mineral, Rio Grande, and Saguache) comprise more than 5.2 million acres with 57% federal land, 39% private land, and 4% State government land. Patterns of land ownership vary greatly across the Valley's counties. Costilla County is 99.9% privately owned whereas Mineral County is 94% federal land. Alamosa County is 61% privately owned with 27% federally owned.

The US Forest Service (USFS) of the US Department of Agriculture (USDA) manages 40% and the Bureau of Land Management (BLM—of the US Department of the Interior) manages 12% of the lands in the Valley. The USFS and BLM operate a combined field office there. Saguache and Mineral Counties have more National Forest land and Saguache and Conejos have the highest concentrations of BLM land. In addition, the Great Sand Dunes National Park and other park lands, managed by the National Park Service, comprise 3% of the Valley and the US Fish and Wildlife Service holds 2.2% of the land. About 2.1 million acres cover the area defined as "the Valley floor."[25]

Many Native American groups believe that the Valley is the source of life where humans and spirits enter and leave this world. The oldest evidence of humans in the San Luis Valley dates back an estimated 11,000 years. The area contains some of the oldest prehistoric archeological sites in North America. The first people to enter the Valley were nomadic hunters of mammoths and prehistoric bison who grazed there.[26] The Capote band of the Ute Confederation inhabited the SLV in Colorado near the headwaters of the Rio Grande; this band makes up part of the present-day Southern Utes with headquarter in Ignacio, Colorado. The Utes hunted for deer, elk, antelope, and other game (using Clovis and Folsom arrow points) and gathered seeds, wild berries, and fruits. They were also known to have planted and harvested corn, beans, and squash. The Utes called the Valley "Tavi-we-a-got" or Big Valley.

Travel to and from the Valley from New Mexico followed the route of today's Highway 285. Although the Valley had been visited by Spanish and American explorers, the Utes and Navajos discouraged settlement of the Valley until the 1840s. In 1842, Mexico established land grants in the Valley, including the Trinchera, Baca Grande, and Sangre de Cristo.

Families occupied the land under the Spanish land grants but by feudal tenure and not in fee; thus, the title remained with the Spanish crown, which has clouded the land-grant issue until very recently.[27]

In 1848, the SLV was ceded to the United States by Mexico in the treaty of Guadalupe Hidalgo, and in 1851 Fort Massachusetts (now Fort Garland), was established. San Luis, the oldest town, was founded in 1852 as the first permanent settlement. The San Luis Valley became part of the Colorado Territory on February 25, 1861. Congress felt that as much land area as possible should be north of the slave-free dividing line. The United States placed most of the Conejos land grant into the Rio Grande National Forest.

From 1878-1880, Mormon settlers established the towns of Manassa, Sanford, and Rome. In 1878, the Denver and Rio Grande Railroad completed a line to Alamosa, thus enabling the export of local products. Alamosa, the major town, is 239 miles from Denver and 141 miles from Santa Fe.

Any water needed for CSP development is likely to be a sensitive issue in the Valley. The Valley receives a scant 6.5 inches of precipitation a year, and its residents are particularly sensitive to water issues. The first recorded water rights in Colorado were filed in this area. The oldest water right in Colorado is attributed to the San Luis People's Ditch, with a priority date of April 10, 1852, in the amount of 21 cubic feet per second from Culebra Creek in Costilla County.The total population of the SLV in 2006 was 48,291 with a density of 5.9 people per square mile, compared with a Colorado population of nearly 5 million and a land density of 46.4 people per square mile. Annual average population growth rates show the region is expected to increase 0.6% per year until 2011; subsequently, the population growth is projected at 1.1% per year. Approximately one-third of the population is Spanish-speaking; about 4.2% of residents are noncitizens.[28]

The six Valley counties are among the lowest of Colorado's 63 counties in median income, costof-living, and purchasing power. Economic activity historically included wheat and barley production, and later timber production and gold and silver mining. Ranching came to dominate the Valley economy by the 1970s. Today the SLV—the highest and largest alpine valley in the world—is the largest producer of potatoes in the United States. The sun shines an average of 320 days per year, and the nights are cool.

The land on the Valley floor is flat. The combination of sunshine, topography, cool temperatures, high altitude, and scant rainfall combine to yield a premier site for CSP.

1.4. Parts of the Report

This report has three parts, which begin after the following section on Methods.

- *Part One: Interview Findings*. This part contains a preliminary analysis of the qualitative stakeholder interviewing completed from October 2008 through March 2009.
- *Part Two: Synopsis of the Transmission Controversy*. After the interviewing was completed, a controversy erupted over a new transmission line to be built from the Calumet Substation into the San Luis Valley. This part of the report contains a synopsis of the controversy that has engulfed the proposed siting of a transmission

line to carry electricity out of the SLV to load centers on the Front Range of Colorado.
- Part Three: Discussion, Conclusions, and Recommendations. This part discusses conclusions from the study and makes state policy recommendations.

In addition, there are two appendixes.

- *Appendix A: The BLM PEIS Process.* Appendix A contains a summary of the BLM Programmatic Environmental Impact Statement (PEIS) process, including excerpts from public comments.
- *Appendix B: Chronology of Key Events in the Transmission Controversy.* Appendix B contains a table showing a brief summary of key events in chronological order.

2. METHODS

2.1. Introduction

This study focused on stakeholder views of a hypothetical 100-MW parabolic trough CSP plant to be sited somewhere in the SLV. The plant would generate electricity from a steam turbine and could be cooled by water or air, although the latter would be more expensive. It would have storage in the form of molten salts, which would keep the power coming well after sunset into late evening hours. Qualitative semi-structured interviews with probes for specificity and completeness were used to collect data from a purposive sample of stakeholders within and outside the Valley.The sample was systematically selected to represent a range of stakeholder group perspectives on CSP facility and transmission siting. Interviewees were asked to reflect the views of their groups.

2.2. The Research Questions

The study set out to identify social, economic, and policy factors that facilitate and impede the siting of a 100-MW CSP in the San Luis Valley. Although considerable research exists on public opinion about energy, which documents the persistent preference of the US public for energy efficiency and renewable energy sources, little research has focused on perceptions of specific renewable energy facilities and renewable energy facility siting issues. Yet, it cannot be taken for granted that the public and, more specifically, stakeholders, will accept the siting of renewable energy facilities. The Cape Wind controversy over the siting of an offshore wind facility near Cape Cod[29] is illustrative of the significance of the stakeholder dimensions for impeding the siting of wind power, and the Medicine Lake controversy[30] shows the difficulty of siting a geothermal facility on or near culturally sensitive lands.

At the study's outset, the investigators met with a Research Advisory Committee of key stakeholders within and outside the Valley to identify key research questions for the study. As might be expected, the stakeholders had a variety of views about and questions for the research. These included the following:

- What are the key governmental entities involved in CSP decision-making and how can they work together?
- Should there be local ownership of a CSP facility or should it be owned by CSP developers? What is the appropriate role of the local players?
- What are new models for siting-decision processes rather than the producer-buyer-purchase power agreement model? Can rural electric cooperatives play a larger role and, if so, how? Is there a model for developing power locally and using it where it is produced?
- What are the key water, land, and other issues?
- What is the level of awareness and knowledge of key SLV stakeholders?
- What is the level of community support and opposition?
- What are the issues relative to transmission, such as crossing federal lands, environmental impact statements required, and public participation?
- What are the policy, regulatory, and sociological barriers to CSP development?
- What are the public lands involved in the Valley and what are the implications of CSP development on public lands? Is CSP included in the BLM's resource management plans? Are there concerns about sensitive Wilderness areas?
- What role are environmentalists playing?
- What are the institutional barriers to CSP siting?

This report documents a portion of the analyses being conducted as part of the Joint CU-NREL study.[31]

Three different approaches were used in this study: (1) stakeholder analysis, (2) review of the BLM's PEIS on solar development in six Western states, and (3) preliminary analysis of transmission line controversy.

2.3. Stakeholder Analysis

In-depth focused interviews with stakeholders relative to CSP were conducted in cooperation with the University of Colorado. Initially, relevant social science literature was reviewed on topics related to facility siting of renewable energy facilities, NIMBY and LULU concepts, facility siting controversies, and political and community factors affecting controversies surrounding facility siting. Then, relying on guidance from a Research Advisory Committee comprised of representatives of key stakeholder groups, and objective methods for identifying stakeholders, we identified key stakeholder group categories for the San Luis Valley relative to CSP.

The key stakeholder groups represented in the study included the following.

- Education, advocacy, and professional organizations within and beyond the Valley
- Agricultural groups (farmers, ranchers, landholders, and their organizations and associations)
- SLV economic groups
- Government and regulatory agencies, organizations, and representatives
- Solar/CSP industry representatives

- Environmental organizations
- Utilities.

It was expected that the potential benefits of CSP development would be perceived differentially by various stakeholder groups, and indeed, could be perceived differently by stakeholders within and beyond the SLV. [32] It was also expected that positions toward CSP development would vary by organizational positions in functional and geopolitical domains. So, for example, an organization such as the local cattlemen's association would look at CSP development through the lens of ranchers potentially concerned about land values and water availability, whereas an organization such as a local hotel owner might be thinking of the profit to be made from "heads and beds" if a CSP project were to go forward.

Anticipating and documenting the expected diversity in perspectives of social groups, as well as finding common ground among them, can strengthen decision processes by providing objective data to support those responsible in working through the multiplicity of interests, power hierarchies, knowledge levels, beliefs, and opinions in a project area. Such "working through" increases the inclusiveness of the decision-making process, potentially increasing the sophistication and effectiveness of decisions. In addition, juxtaposing local views and local expertise enables the community itself to more accurately construct the complex and colorful tapestry of its own positions, beliefs about impacts, and views on potential benefits and costs. Outside stakeholders bring considerable knowledge to consideration of the impacts of CSP development. Increasing the awareness on the part of those organizations *outside* the Valley, in positions of decision responsibility, about their own interests and how they relate to those *within* the community of the proposed siting contributes "social facts" to illuminate the complex problem of utility-scale solar development.

2.3.1. Sampling

The interviews were conducted with a purposive sample. Because of the limited resources available for the project, the stakeholders identified were divided into tiers by type of stakeholder group and the first-tier stakeholders were approached for interviews. Interviewees were sent letters explaining the study to them and informing them that they had been identified as part of the study's sample. Most of the sample agreed to be interviewed; however, some did not. In those cases, a back-up organization in that same category was approached. Probably because of tensions already mounting about the siting of transmission lines in the Valley, a few stakeholders both favorable and unfavorable to CSP development (positions especially seeming to relate to the proposed transmission lines) failed to respond or declined to be interviewed. Nevertheless, the range of stakeholders interviewed was broad and we were able to complete interviews with others in the same stakeholder group.

The interviews were conducted with representatives of the organizations to which they belonged. Therefore, the perspectives given represented the stakeholder groups for which interviewees spoke, not their individual attitudes, opinions and views. Between October 2008 and March 2009, 25 interviews were completed with organizational representatives. Fourteen of the interviews were conducted with residents of the San Luis Valley and 11 with knowledgeable stakeholders who resided outside the San Luis Valley. Most, but not all, of the organizations represented in the sample were located in Colorado. One limitation of the study is that no interviews were completed with representatives of the Valley's Spanish land-grant families or Native American tribes. The findings cannot be generalized to the population of the San Luis Valley or even to all potential stakeholders in the Valley. The findings provide descriptions of the key issues and stakeholders—as defined by some of the organizations who know the most— about the technology, other stakeholders, the decision processes in siting a CSP facility and the proposed transmission line, and the community where this would be done.

To appropriately appreciate the interview findings represented in this report, one needs to understand the starting positions of the organizations relative to CSP development in the SLV. Within the six-county Valley community, organizational representatives knowledgeable about their own and others' interests were included.

The Valley interviewees varied in their knowledge levels about utility-scale solar development, although everyone interviewed knew about the 8.2- MW SunEdison Solar PV plant sited on Highway 17 near Mosca, and their positive experience with it tended to color their responses toward CSP. But if not all were expert on CSP, the Valley stakeholders were expert on the Valley, its culture, its residents, its economic situation, and its politics. In addition, some of them favored CSP development and others did not; therefore, it was possible to capture a variety of organizational views toward CSP and transmission development.

Those interviewed *outside* the SLV were highly knowledgeable representatives of organizations with a stake of one kind or another in CSP development in the SLV. To capture the range of relevant stakeholder opinion, those stakeholders were interviewed who, by objective analysis and subjective experience, have an interest in what happens in the Valley in terms of large-scale solar energy development.

Thus, the stakeholders *outside* the Valley tended to be highly knowledgeable about the technology and relevant policy, to be favorable to CSP development, and they embraced claims made for its benefits both inside and outside the SLV. However, they also knew about potential negative impacts and were concerned that these be accurately identified, anticipated, and taken into account in any CSP project planning.

2.3.2. Data Collection

As noted, the interviews were semi-structured, open-ended interviews with extensive probing for specificity and completeness of responses.[33] Questions focused on the stakeholder's definition of the then-current situation in the SLV, the perceived advantages and disadvantages of renewable energy generally and of CSP in particular, the definition of SLV stakeholders with respect to solar energy development, and perceived considerations relative to facility siting and transmission. The interviews ranged from 60 to 90 minutes in length and were tape-recorded and transcribed verbatim, resulting in approximately 400 pages of field notes.

2.3.3. Analysis

These notes were then subjected to content analysis to identify discrete ideas and comments volunteered by stakeholder interviewees. The importance of the voluntariness of response is underscored; no suggested answers were raised in the interviews, so that the responses emanated from the particular stakeholder organizations' own already-held positions. The content analysis was intended to identify the scope, range, and texture of ideas about perceived impacts—positive and negative—from various stakeholder positions. It is possible to identify the significance, in the sense of the "emotional charge" of these issues, as well as to gain insights into what was (at the time of the interviews) beginning to shape up to be a difficult controversy about the transmission lines proposed to export solar electricity out of the Valley to Colorado's Front Range electricity load centers.

2.4. Review of the BLM's PEIS Process

As noted earlier, fostering development of renewable energy is national and state policy. Given the amount of federal land in the SLV, and the importance of the Valley as a site for solar development, it is inevitable that federal policy on solar siting on public lands would be germane to the SLV. By 2008, the BLM had received more than 150 requests nationwide for permits to site large-scale solar facilities on public lands, yet the agency's Resource Management Plans (RMPs) did not include solar development. By law, land-use permits cannot be issued for actions that are not covered in RMPs because environmental and socioeconomic impacts have not yet been assessed for such actions.

During the mid-2000s, federal policy shifted toward fostering utility-scale solar development. To streamline the permitting process for these facilities, the BLM initiated a PEIS process to examine the impacts of solar developments in six Western states on May 28, 2008. In 2009, Secretary of the Interior Ken Salazar established Renewable Energy Coordination offices in Arizona, California, Nevada, and Wyoming, and smaller teams in five other states including Colorado. This followed on the 2008 announcement by BLM and DOE designating 24 tracts totaling 670,000 acres of public lands as Solar Energy Study Areas (SESAs), including 21,000 acres in the SLV. The four sites are DeTilla Gulch, Fourmile East, Los Mogoles East, and Antonito Southeast.

If fully developed, the Colorado sites could generate more than 4,000 MW. Companies proposing projects in these areas would be eligible for faster permitting processes. A PEIS addressing the impacts of solar development in six western states is being completed in order to speed the permitting of facility siting on public lands. The final PEIS and Record of Decision are expected in the fall of 2010. Appendix A of this report provides an overview of the BLM PEIS process, public comments from Colorado organizations and individuals, and its current status.

2.5. The Transmission Controversy

A significant controversy erupted in May 2009 when Tri-State Electric Generation and Transmission Association and Xcel Energy applied to the Public Utilities Commission of Colorado for a Certificate of Public Convenience and Necessity (CPCN) to build a 230-kV transmission line into the SLV from Walsenburg to Alamosa, Colorado, along Highway 160 through the privately owned Trinchera Ranch. Although the stakeholder interviews had yielded perceptions of transmission issues, they were completed before the legal and community opposition to the transmission line crystallized. Therefore, Part Two of the report includes a synopsis of the controversy that developed after the interviews were completed based on documents, news articles, reports, and other artifacts concerning the transmission line controversy. This analysis summarizes a recent history of events leading up to and encompassing the controversy, including identification of the key stakeholders and the issues involved.

PART ONE: FINDINGS FROM THE INTERVIEWS

3. PERCEIVED ADVANTAGES

Summary

Stakeholders identified economic, environmental, technological, social-psychological, and policy and regulatory advantages of renewable energy generally and of CSP development. Patterns of response differed somewhat by whether interviewees came from within or outside the Valley. Valley stakeholders were more articulate and passionate about local economic and socialpsychological/identity values, and those outside the Valley emphasized broad environmental and technological advantages. Valley stakeholders were more concerned about poverty in the Valley and the cost of electricity, availability and persistence of local jobs, and making certain that the Valley's residents would benefit financially.

3.1. Introduction

Stakeholder informants were asked about their organizations' view on the advantages and disadvantages of renewable energy generally and of CSP in the San Luis Valley.

The 396 pages of text were content-analyzed to discern discrete mentions of perceived advantages of renewable energy generally in the context of open-ended questioning; these are considered "responses." Five broad categories of advantages were developed from the comments themselves: (1) economic, (2) environmental, (3) technological, (4) social-psychological, and (5) policy and regulatory. Under each of these broad categories, subcategories were defined. Each response was coded by category and subcategory.

Table 1 shows that 231 discrete responses mentioned advantages of renewable energy generally and CSP specifically were identified.

The table is organized in terms of decreasing frequency of mention (or responses) of the different types of advantages, with economic advantages receiving the most emphasis in the data set overall (44%), followed by environmental advantages (26%), technological advantages (19%), social-psychological advantages (9%), and policy and regulatory advantages (2%).

Overall economic advantages are mentioned far more frequently than are environmental or other types of advantages by both types of stakeholders. Stakeholder interviewees inside the SLV mentioned economic advantages of CSP specifically in the Valley almost twice as often as they mentioned economic advantages of renewable energy generally, suggesting that local economic advantages of a proposed CSP facility weighs more in locals' positions toward the facility than other considerations.

Stakeholders outside the Valley more often mentioned "technological" advantages than did those in the Valley, suggesting their higher level of technical sophistication about CSP.

Third, environmental advantages of both renewable energy generally and of CSP specifically were mentioned somewhat more frequently by Valley stakeholders than by outsiders.

Fourth, Valley stakeholders mentioned "social-psychological" advantages more frequently than did those outside the Valley. Each of these the categories of advantages is discussed and exemplified in more detail below.

Table 1. Types and Percentage Distribution of Responses about Perceived Advantages from Stakeholders within and outside of the San Luis Valley (n=231 comments)

Types of Advantages	Renewable Energy Generally		CSP Specifically		
	% Responses Within SLV	% Outside SLV	% Responses Within SLV	% Outside SLV	% Totals
Economic	27	38	53	52	44
Environmental	40	34	20	15	26
Technological	25	15	11	26	19
Social-psychological	4	6	16	7	9
Policy and regulatory	4	7	--	--	2
	100% (n=48)	100% (n=53)	100% (n=64)	100% (n=66)	100% (n=231)

3.2. Economic Advantages

Economic advantages identified tended to focus on the potential economic impacts of CSP within the Valley, rather than on effects of renewable energy generally.

The 44% of discrete responses categorized as economic (n=67) fell into six somewhat overlapping categories of advantages, discussed below.

Most frequently mentioned were *potential improvements to the local SLV economy* from the siting of a CSP facility there.

Two quotes from Valley stakeholders:

> ...it'll happen. For everyone [but farmers], this is a no-brainer. If we just had the wherewithal as a little rural community to make this happen and to be sure that the SLV reaps the rewards...how can we make sure that the SLV reaps the rewards? Because someone could build a solar plant and all the money leaves the local economy...we'll, we've done good for our planet, but how can we make this benefit our own economic development?

> ...our intent to have this valley 100% renewable by the year 2010. And that could happen in an hour if Tri-State, Xcel, and Rural Electric all agreed that they could get mutual benefit because we could actually purchase enough [renewable energy credits] to have the Valley 100% renewable based on existing energy...

Proportionally more stakeholders outside the Valley than within it referred to economic benefits for the people living in the SLV; it was one of several benefits of CSP that they cited to describe the benefits of siting a CSP facility there.

...the new energy economy has been a huge boom to our state with respect to job creation and economic development at a time when we really need it the most.

They use a local resource so they create economic activity in the Valley... It adds another engine to the Valley's economy. The Valley is pretty much agricultural as I understand it, so you bring in a very clean other business.

having the ability to create a new sustainable energy industry, also looking at the opportunity that we could generate as much as $2 billion in new investment in rural communities which are so desperately in need of....

On the other hand, Valley stakeholders stressed the idea that *renewable energy generally creates jobs and that a local CSP facility could lead to the location of a manufacturing facility in the Valley.*

This notion was circulating fairly widely within the Valley at the time of the interviews, and appeared to be based on statements made by the CSP industry.

And it would create a large number of construction jobs—about 1,000-2,000 over a two-year period) and a smaller number of maintenance jobs, say 60 to 70 full-time jobs—permanent, good-paying jobs for the 30- to 50-year life of the plant....

A local informant said:

It's clean, we have a lot of sun, there are no negative effects and it makes some jobs—and there may be manufacturing jobs—if this were a big site for solar, it could lead to the manufacturing of solar components. That would improve the economy. It would be good for some businesses—contracting with them to place on rooftops, power their business and sell excess to the SLV REC [Rural Energy Cooperative] or Xcel. It's an advantage for business.

Another economic advantage of renewable energy generally seen as crucial (but not specifically mentioned as an advantage of CSP) is that it could *stabilize energy prices* at a time of rapidly increasing prices of coal and natural gas.

Fuel price stability would probably be the single most important facet with respect to CSP...the Governor's Climate Action Plan...calls for a 20% reduction in statewide greenhouse gas emissions by 2020 and an 80% reduction by 2050....it is important to note [that] the current Climate Action Plan models a 200-MW concentrated solar power facility somewhere in Colorado.

Two other economic advantages were mentioned: (1) *providing tax benefits to local counties* (which was mentioned more frequently by stakeholders outside the Valley [six mentions] than by ones within the Valley [one mention]) and (2) *trapping the dollars in the state's economy* that are now sent out of Colorado.

In fact, one Valley interviewee commented that there should be a severance tax for exporting solar electricity from the Valley.

Do you recall the Federal Mineral Lease, or the Severance Tax, things that are going on with all of that? Did you hear the Western slope and Grand Junction and that area talk about their severing the minerals from our...we should get some of the benefit of that here, you've heard them say that?
Interviewer: Yes.

Stakeholder: While solar is not a severable mineral, locating your things on our vistas is use of our resource. Should there be a severance tax and could that tax benefit these communities?

Interviewer: For economic development?

Stakeholder: Yeah, we've got the poorest school districts in the state, I mean on and on. Could the social gain be...we have this resource here, and people from the outside want it, they want our sunshine that's produced in our valley, could we tax for their use of our sunshine to benefit communities located here?"

Although the supporters of CSP development would probably not agree, the prospect of extra local tax benefits tailored to the solar resource combined with the clear air and flat land that would help mitigate the poverty in the San Luis Valley would very likely be welcome to the Valley community.

3.3. Environmental Advantages

More than a quarter of the responses about the advantages (26%) addressed environmental advantages (n=60), mostly of renewable energy generally.

Most frequently mentioned were that renewable energy, and CSP specifically, puts excellent San Luis Valley solar and natural resources, including lands, to best use (25 responses). Following are quotes from SLV interviewees.

We have 8,000 square miles, a lot of which is chico brush and land that is not being utilized, so there is plenty of room here.
I think the land owners who have owned....it's just not good farmland. It's chico brush and either maybe they don't own the water or the soil is not right because there is a lot of alkaline in the valley....so, like where they put that SunEdison...maybe that is the highest best use for that. So the people who sell that land, that may be their only opportunity to sell that land.
If you do some due diligence you can pick areas of the Valley that are fairly rocky, areas were farming is not done because soil isn't good for farming potatoes. Other areas aren't producing much of anything—those areas would be the priority areas for solar panels. [The interviewee was talking about the 8.2 MW Sun Edison
concentrating PV plant—this confusion between CSP and the PV plant existed among several SLV stakeholders.]

Another response was:

If you've got property that received the same amount of sun as the irrigated acreage that provides potatoes for food, you wouldn't take one in favor of losing the other, but you would select the best one for the use... I suppose that says this makes more sense to sort of take the crops out of projection if you choose fallow land. It's the highest cash use.

A second type of environmental advantage cited (15 mentions) is that renewable energy *reduces greenhouse gas emissions, addresses climate change, and helps to "save the planet."*

It can probably be assumed that this type of advantage is thought to apply to CSP as well, although no responses cited this advantage relative to CSP in the Valley specifically. Quotes from within the SLV exemplifying this type of comment are as follows:

Whether you believe in climate change or not, the climate's different than it used to be for whatever reason. The fact is that we need to address that. I think renewable energy is important [in addressing climate change].

The biggest advantage is that it is renewable, not a finite source of energy. Reduction of greenhouse gas emissions is a benefit to renewable energy development—carbon reduction.

Quotes from outside the Valley on this point include:

I think in order of importance, probably climate change and greenhouse gas emission reduction targets, secondarily the economic development piece, and perhaps third the fuel price stability and long-term cost reduction.

Quantified - 1 billion tons of carbon reductions.

The third type of environmental advantage is the reference to renewable energy as *"clean energy"* and *"good for the environment."* Thirteen discrete responses, mostly from SLV interviewees, referred to clean energy, but only one relative to CSP.

Example quotes are as follows:

Renewable energy is good simply because it is renewable...it is popular to develop renewable energy because of environmental reasons, like clean air and water.

Other environmental advantages identified were that *renewable energy uses less water* than other forms of energy (two responses—one inside and the other outside the Valley), that it *helps wildlife* (two SLV responses), and that it *won't disrupt the vistas* (two outside responses).

3.4. Technological Advantages

Emerging from the content analysis was a set of stated advantages that were classified as "technological" advantages, comprising 19% of the responses.

SLV stakeholders mentioned at least 17 of these and stakeholders outside the Valley mentioned 25 such responses.

Twenty responses mentioned that renewable energy *improves the security and diversity of power supplies and reduces dependence on imported energy.* One Valley stakeholder put it this way:

> [What's important] is the ability to attach and I think what it is called is...they've got to be able to integrate with our system through a substation to get it out, and right now there is a 230, actually it's a double-circuit, 230 transmission line planned in *the San Luis Valley to Walsenburg, and it is a combination of Xcel and Rural Electric and Tri-State that are going to put that transmission line in.*
>
> *Interviewer: Is that for sure going to...?*
>
> *Stakeholder: Oh, yeah. One way or another. And it should be....the SLV doesn't put any money in, it is just for the Valley system, and Xcel and Tri-State pay for it. It is about an $80 million project. It should be completed by 2012 or 2013.*
>
> *Interviewer: And why are they doing this?*
>
> *Stakeholder: Because we are running out of power in the valley. The last transmission line built in San Luis Valley was 35 years ago.*

One stakeholder from outside the Valley described the advantages of renewable energy in a different way.

This interviewee stressed the security value of distributed generation, as shown in the following quote:

> We talk about what we're doing today as the alternative model and that what we'd be doing in the future is going to something traditional, I would say the centralized model is the alternative model. For the life of the planet, we have dealt with distributed generation and energy for all but about 70 years of the life of this planet.
>
> Integration of the distributed and larger grid can complement the central system with energy efficiency and renewable energy, instead of having another 25 or 30 coal plants in the queue in this country; it should be classified as preposterous, when we know that with energy efficiency, conservation, and the distributed model, we can take this centralized system that we've built and is in play and is working well for us, and complementing it and meeting our future demands with a much more distributed model.
>
> It's the smart grid that is the glue or provides the way to accommodate distributed and central station. Once we got the next advances in storage, then we really have a system...
>
> Advantages of...
> It's more resilient. If you look at it from a national security perspective, and you quantify and weigh national security, economic security, and environmental benefits, this model is what maximizes those three things.

A well-informed stakeholder from outside the Valley talked about the prospect of CSP development and the need for a transmission line to enhance the security of the Valley, which has been served by only one transmission line coming in from the north.

> It was a cooperative transmission line across La Veta Pass into the Valley. It goes across some private land, some state land, and then it also might go partially along a highway corridor, so you're bringing it across the pass. What it is is it is a double, I think it is a 230-KV line that will be a shared project from Tri-State and Xcel. One, they have to have a loop system right now into this valley. Their energy into the valley is very insecure because if there were a fire, these guys are toast. There wouldn't be any electricity for quite a while because they are very vulnerable the way the whole transmission network is in the San Luis Valley.... it's for the security of the region.

Another outside stakeholder put it this way:

> It's flipped. Energy independence probably takes precedence over clean energy right now, and that's the whole Middle East thing. So the priorities kind of swapped. In fact if you listen to Obama's speech he talks about eliminating our dependence on Middle Eastern oil. I mean he says Middle Eastern which is key, by, in ten years, and that's all security.

Clearly, the energy security implications of renewable energy were described from differing perspectives, but similar basic themes emerged about increasing US self-sufficiency through decreasing foreign imports, and increasing Colorado self-sufficiency by producing Colorado power (not importing coal, for example).

A second type of technological advantage was classified as *CSP provides dispatchable energy* and is a *building block toward baseload electricity*. Outside stakeholders (n=13) made more comments about this than did Valley stakeholders (n=5).

> CSP has two distinct advantages over PV right now. I think costs are roughly equivalent on an energy basis. However, CSP does not suffer the short-term intermittency that PV does.... with the SunEdison facility, there is more volatility with generation...CSP avoids this because it is a thermal cycle. It is essentially a steam turbine and it does not react to individual clouds the way PV does, and it has the ability to add storage inherent in the system . . . the advantage of thermal storage is that with 100% reliability it meets peak day loads.

3.5. Social-Psychological Advantages

One of the most intriguing perceived advantages identified by Valley stakeholders was the discovery that they see the potential for social-psychological advantages.

A verbatim sequence of one of the SLV stakeholder interviews illustrates this point:

> The sun is readily abundant in the valley. This is the land of cool sunshine. It's colder than heck in the winter but we do get a lot of sun, so...there are actually some advantages I understand from being at this altitude and this climate for concentrated solar because we don't overheat such as you would say in Death Valley. That's my understanding, I don't

know if that is totally accurate, but I see that as a benefit. I think....and this is a real intangible, I don't know if you know this, but the San Luis Valley has five of the poorest counties in the state and this is something to be proud of.

Interviewer: Something cutting-edge....

Stakeholder: Yeah, and, well, it puts a different face on the San Luis Valley as far as people on the Front Range come through here and don't understand the beauty of the place because either you love it here or you hate it, but a lot of people from the metro areas are like, oh my God, how can you live in a place like this, and a lot of people who live here don't appreciate what they have, so I think changing our perception of ourselves....I mean this could be....and that's why I have a concern about....if they come and build everything and then take everything out of the valley, if they could just come and build everything and share some of that...it's kind of like Los Alamos, you're familiar with Los Alamos? Well, it has the highest concentration of Ph.D.'s in the country, so if some of that technology comes here and some of the people, some of the subject matter experts, and some of the industry associated with it, for example the manufacturing of the actual...whether it's instrumentation, maybe it's just like a silicon chip, I don't know what it is, if they bring industry with it that is associated with it and turn this into....it can't be the Silicon Valley, but more value-added than putting the stuff here and then taking everything out.

Interviewer: ...somehow there is a local involvement in that there is some kind of manufacturing capability and interest that stays in the valley. Is that the idea?

Stakeholder: Yes. That they're committed...that they become a valued business and they are committed to the sustainability of our valley. I don't think the majority of the citizens in this valley want to change. We don't want to be another metro center. We don't want to be an Aspen or a...we'll never be that, but we want to sustain the quality of life...
Interviewer: Why not?

Stakeholder: Because the quality of life that is here is that we are not Aspen, we're not Vail, we're not Denver, we're not Boulder—that's the quality that needs to be embraced by a business that comes here and understood by a business that comes here. One of the unique things—and this is part of our mission, to enhance and maintain the unique culture and heritage of the Valley because of the strong Hispanic influence, the Native American influence, the Mormon influence—the history of the valley needs to be understood by companies that come because the majority of the people who live here like it the way it is.

Another Valley stakeholder put it this way:

If you look at demographics of the Mississippi River Delta that compete with the San Luis Valley for a poor economic situation, the valley itself is a distressed economic area. What would be the impact to the rest of the world, the rest of the United States, the rest of Colorado, if the poor San Luis Valley pulled itself up by the bootstraps to where it becomes the first renewable valley in America?

> There would be entire vital reach in the ecosystem, the whole dynamic of its own future is suddenly 100% renewable energy. Now that's an opportunity to erase at least one level of myth of what you are as opposed to what you aren't.
> It is a way for people to socially elevate themselves to be able to walk into any store....
>
> *Interviewer:* A source of pride.
> *Stakeholder:* It is a source of pride. I mean a little of it is that I don't like the rah-rah, go purple and white, fight-fight story...but there are aspects about being able to tell folks that with existing infrastructure, existing people, existing talent, existing local money, that you became the first 100% renewable bio-region in the nation, that the headwaters of the Rio Grande is 100% renewable.

3.6. Policy and Regulatory Advantages

The final category of perceived advantages was the recognition that the development of renewable energy generally helps Xcel Energy meet its renewable portfolio standards (RPS) requirements, with two responses from the Valley and four from outside the Valley on this point. An illustrative quote from outside the Valley:

> I think Amendment 37 was the primary reason that [Xcel Energy] started acquiring solar resources...Amendment 37 tells each Qualified Retail Utility in Colorado how much of its retail sales have to come from renewable energies, and there is a carve-out within that for solar technologies subject to a retail rate impact cap. If it was going to cost too much money to acquire those resources, then the utility doesn't have to meet the energy requirements.

Amendment 37 was the first citizen-initiated, statewide RPS in the nation; many SLV citizens support the RPS and want it to be met, but did not mention specific advantages to the Valley from it.

4. Perceived Disadvantages

Summary

Stakeholders identified potential disadvantages of CSP development. Interestingly, the most frequently occurring comments on disadvantages pertained to environmental issues, especially concerning water constraints and the intensive use of land that CSP development would entail. There were also some concerns about wildlife, weeds growing under collectors, and the idea that crops could not be grown under collectors. In addition, there were many more Valley comments (n=23) than outside comments (n=8) on the economic disadvantages associated with CSP development, including high front-end costs and the potential for a boom-bust cycle in the local economy. Other disadvantages remarked on, by both Valley and outside stakeholders, related to technological aspects (such as intermittency and need for transmission), but more comments from Valley than from outside stakeholders focused on social-psychological and policy and regulatory disadvantages of CSP development.

4.1. Introduction

The field notes were also content-analyzed to discern discrete mentions of perceived disadvantages of renewable energy generally and of CSP development specifically.

The responses on disadvantages lent themselves to categorization in the same categories as the responses on perceived advantages: (1) economic, (2) environmental, (3) technological, (4) social-psychological and (5) policy and regulatory. Again, under each of these broad categories, subcategories were defined. Each response was coded by category and subcategory.

Table 2 shows that 158 responses mentioned disadvantages of renewable energy generally and CSP specifically. The table is organized in terms of decreasing frequency of mention (or responses) of the different types of disadvantages, with environmental disadvantages receiving the most emphasis in the data set overall (42% of the responses), followed by economic disadvantages (23%), technological disadvantages (19%), social-psychological disadvantages (11%), and policy and regulatory disadvantages (4%). One outside stakeholder said there were no disadvantages of CSP siting at all.

Comments on environmental disadvantages occurred more frequently in the interviews than other types of disadvantages and represented the most important area of concern for both types of interviewees (within Valley n=37; outside Valley n=30). Responses from the Valley cited economic disadvantages for renewable energy generally and CSP development specifically (n=26) more frequently than responses from outside the Valley (n=10).

Table 2. Types and Percentage Distribution of Responses about Perceived Disadvantages from Stakeholders within and outside of the San Luis Valley (n=158 comments)

Types of Disadvantages	Renewable Energy Generally		CSP Specifically		
	% Responses Within SLV	% Outside SLV	% Responses Within SLV	% Outside SLV	% Totals
Environmental	24	40	41	48	42
Economic	38	13	27	16	23
Technological	38	20	13	26	19
Social-psychological	--	20	13	8	11
Policy and regulatory	--	7	6	2	4
	100% (n=8)	100% (n=15)	100% (n=85)	100% (n=50)	100% (n=158)

4.2. Environmental Disadvantages

The most frequently mentioned disadvantages of CSP development were environmental issues.

The two major perceived environmental disadvantages are (1) water constraints and the (2) land-use intensity of CSP.

4.2.1. Water Constraints

The most frequently occurring environmental comments concerned water availability by stakeholders within (n=11) and outside the Valley (n=12). The following two quotes are illustrative.

> CSP is problematic. It is very problematic in terms of the region's ability to conceptualize what it is going to look like and what it is going to take, that water is paramount. You have a hydrology unlike any other place in the world and you have citizens in this region who are so committed to their water and using their water carefully, using this water...it's a closed basin...aquifer, part of it is a closed basin and then above it is an open aquifer that is what provides the water for the irrigation. It's an extraordinary bit of hydrology here...it's the most contentious issue here. That's the problem with CSP. It requires water. You have some of the most unbelievable water [battles over water]...these are people who are more incented by their environment, their air quality, their water quality, and their quality of life than they are incentives by having industry come in...maybe that would change if they got a taste of it, but right now, anything that threatens their water has them quite concerned.
>
> Developers need to secure access to water—it depends on cooling water, cooling tower—evaporates water to cool it. CSP needs 700 gallons of water per MWh. This could be an issue in the SLV, but it's not a show-stopper. Fans can be used to air dry the plant, but they consume electricity; this would be an economic decision based on the cost of water, the availability of water, and a pro forma financial model. Technically, at least, it's feasible...everybody knows the SLV is an ideal location; farming is on the decline because the water table is lower; people are trying to figure out how to make money. They want to tie up both land and water...water is in shorter supply than the land.

Another outside stakeholder said:

> You shouldn't do CSP without water (just like you shouldn't do coal without water), because it ruins the economics.

The integrated analysis of energy and water resources is gaining more attention in research circles as it is increasingly understood that developing energy resources depends on water, and developing water resources depends on energy.[34] Water has historically been and remains a contentious issue in the Valley, and knowledgeable interviewees acknowledged this. Indeed, a stakeholder analysis of the correlation of CSP and water stakeholders and issues in the SLV should be the subject of separate analyses, and other analyses are needed on the regulatory implications of CSP and water.

Complicated and difficult-to-understand water issues are involved. According to some Valley residents, the SLV has north and south aquifers. Saguache is a closed basin with a three-layered physical structure. At the top layer is surface water; the layer underneath is hardpan; the aquifer lies below. Local residents said that because the layers are relatively impermeable the surface water does not recharge the aquifer. Natural artesian wells spring up from the aquifer through the hardpan. According to some local residents, water is also pumped out to New Mexico to meet water compact agreements.[35] One of the interviewees said that the water priority system is being circumvented by federal subsidies. The highest priority was said to be surface water rights. With more drilling, the surface water was said to be drying up, so the system now is one in which people have to buy and pay for the ability to drill based upon the amount of surface water they own. Based upon their ability to own or buy surface water rights, some irrigating farmers will stay in irrigation and continue pumping. Others will refuse to pay for surface water rights, in which case federal subsidies compensate them for taking their land out of irrigation.

4.2.2. Land-Use Intensity

NREL estimates that CSP technology requires five to ten acres of land per one MW of electric capacity.[36] If this is the case, then 500 acres to 1,000 acres of land would be needed for a 100- MW facility. To be cost-effective, a CSP plant has to be a minimum of 100 MW in size and is probably limited at 250-300 MW per facility. One Valley stakeholder said that one of the CSP developers had said publicly that the company was looking at 5,000 contiguous acres on which to site a facility.

Following are comments from the interviews on the perceived *land-use intensity* of CSP.

> Solar still takes up space in our area.
> [Solar energy] is an intensive use of land...
> ...footprint is a disadvantage in general for renewables...the footprint interacts with habitat loss because it requires such a large footprint, so that it's a significant concern.
> There are rumors out there. Large corporations from California, from Australia, from I think I've heard Europe, that are talking to groups of people here about acquiring land for concentrated solar.... It's a small community so you hear all of this stuff going on. You don't know what to believe.

Another point a stakeholder made is that there are sensitive wilderness areas in the San Luis Valley. The example mentioned was Saddleback Mountain, reaching 7,956 feet, located in Conejos County and used for hiking.

Developers could see threat in land speculators who are driving up the cost of land or tying up the land. Developers could see threat in "locking up" land to prevent development of CSP projects, although this information must be closely held in the Valley as interviewees did not mention any specifics.

Stakeholders varied in defining how much land the Valley comprises. One said that 5,000 acres is one-third of the square footage of the Valley. Another said that 95% of the land is federally managed. A third mentioned that each of the six counties in the SLV has a different pattern of land ownership. Another stakeholder said, "...well there are about 5 million acres in the San Luis Valley, and people say, you're going to use 5,000 acres with solar panels, and I say, but you have 5 million acres."

4.3. Economic Disadvantages

The most frequently mentioned economic disadvantage of CSP development was that a *boom-bust cycle* could result. The cycle would involve the hiring of workers to build the facility with the concomitant economic impacts such as increased demand for housing and increased prices in general, crowding, and demand on public services, followed by a decline in the number of jobs once the facility is built.

This boom-bust cycle in energy development has been well-documented in the literature.[37]

One comment from an interviewee outside the Valley related to this is as follows:

> I think the biggest concern for local residents is something that can happen on a national basis anywhere, whether it is a wind development or a coal plant or a Wal-Mart, is that locals want to know that they will be benefitting, that there is a long-term benefit to them, that there are potential employment opportunities, and that essentially a company is not going to come in and develop a project and then abscond once the rewards have been taken, that the equipment will be renewed and so forth and so on.

Many Valley interviewees used the 8.2-MW SunEdison concentrated PV facility near Mosca as a point of reference in discussing the expected impacts of a potential CSP facility.[38] One said:

> They had temporary jobs to install the solar facility. There were 35 to 38 people hired. It paid well. There were 300 at the job fair, and there was a lot of interest because it was good pay—$14 to more than $20 per hour.
>
> These were laborers, electricians, welders, and equipment operators. [The only job since the plant was built] is the main supervisor. I don't know of any continuous employment from the SunEdison plant.

This interviewee said that when the oil and gas industry came into Parachute, Colorado, they created a boom of jobs paying good money, so the rent on homes there had been increasing.

Then the local people couldn't afford the higher rent, he said, so although it was good for businesses like hotels while the boom was going on, the overall economic situation was not good for the local residents. The boom cycle was not sustainable.

4.4. Technological Disadvantages

Responses in this category stressed the problem that, for a large CSP facility to be sited in the Valley, *transmission is needed* because the Valley is remote from electricity load centers. Three illustrative quotes follow.

> One disadvantage is that the electric transmission infrastructure in the Valley is severely limited so infrastructure would have to be built.
>
> The challenge of connecting to the existing grid...
>
> The direct drawback would be how are you going to export it? It goes back to bringing in new transmission lines.

Another technological disadvantage was that the *efficacy of CSP was still in question*. CSP was perceived by some not to be a proven technology, not yet ready to be commercialized, and that it will take time for it to be a reliable technology. Two illustrative quotes follow.

> With the solar thermal with storage carve-out, it was an attempt to push that technology along, to see how it works in Colorado. Colorado has a good solar resource but we've got extreme winter cold temperatures and we have got concerns about how solar thermal technologies will perform in those extreme winter conditions.
>
> Current solar thermal plants have to be maybe 100 MW in size before they become cost-efficient.

In addition, some comments from within and beyond the Valley questioned the efficacy of CSP as not yet a proven technology or not yet at commercial status.

For example, one Valley stakeholder said:

> ...at this meeting they said concentrated solar hasn't been proven. That's what [one of the Valley experts] thinks, and his opinion matters...this technology is pretty much all in an experimental phrase. Yeah, some places have had it for awhile, but they are still doing a lot of experimentation.

The *intermittency of solar power* was also mentioned several times (n=12 comments). An outside stakeholder said, for example, that a CSP plant would probably have a lower "availability factor" than a coal plant.... the sun goes down, but the storage will supplement some of that, but if you pushed storage to the point where it was an around-the-clock plant, you'd probably be out of the economics."

4.5. Social-Psychological Disadvantages

Fifteen comments, most by Valley stakeholders, fell into the category termed "social-psychological" disadvantages.

There was concern that a CSP facility would mean inequitable benefits—"short-term benefit for the very few"; CSP would be *difficult to conceptualize* and people would not understand the impacts.

One Valley stakeholder said: "I think they're favorable to solar, but they don't know what concentrated solar is. I would say the majority of people, even a super-majority, don't know what that is."

Another point mentioned several times was the fear that renewable energy is "too green" and that *it could be risky* to allow it to come into the Valley. One interviewee put it this way:

> ...we'll have folks, elected officials, who will say, 'well this concept is too green for me' with the idea that if I have to agree with every environmental condition of every group that comes in the door, so there is almost a piece that if I let down the guard protecting the history of how we've done things in the past and instead pick up the idea that by unlocking the door, I'll get more....
>
> If you decide too fast, you can be judged the fool, but if you judge too slow it is not often that you become the hero ... and we are trying to answer the question, is it safe?
>
> Not the technology side, but that they would be judged as making inappropriate decisions...there's the risk. If you say let's go build it and it's financed...but in the end it doesn't produce power or some new technology comes out that is better or if someone changes the laws...

well, they'd say those guys didn't think that through very well.
There is also the sense that it is difficult for the Valley people to deal with Xcel Energy.

> There are collaborations [between the rural electric cooperative and Xcel Energy] as far as the transmission lines, bringing in a pair of 230-kV lines.
> The other part, they focus very differently because of who their ownership is.
> That is one reason that I think the REC is more cautious in getting involved in renewables—their customers are the owners whereas Xcel is much more remote as far as influence one can have on them....it's like the big giant that nobody can talk to, whereas the REC has its ears open.
> The corporate world, when they have their mind made up to do something it is difficult to make changes and sometimes you have to go through some kind of embarrassment such as news media to bring things around a little bit.

This suggests a certain feeling of powerlessness to affect the outcomes of utility-scale energy decision-making.

4.6. Policy and Regulatory Disadvantages

One of the types of disadvantages of regional development of CSP that stakeholders mentioned in the six counties that compose the SLV involved *disjointed county-by-county regulation*.

One Valley stakeholder said:

> It's disjointed now. So you could have one county...Saguache has very lax building restrictions and it's a very poor county, so say if they want to attract solar into their county, that then impacts the surrounding counties because if their incentives are different than other counties...with six different counties in the Valley if you don't coordinate the plan, it is going to be disjointed.

The regulatory situation at the local level may need further development, a coordinated regional approach to be implemented. This could be expensive. Another point is that the BLM has to do a *NEPA analysis*, and that *federal regulations have not caught up* with CSP development.

On the other hand, CSP development on private land requires no NEPA analysis. Finally, the point was made that *the investment tax credit of 30% doesn't help developers* because there is no way to turn it into money.

Although these disadvantages could be counted as barriers or as economic effects, the lack of adequate regulatory capacity would have negative consequences of various kinds for the people of the Valley.

In the meantime, counties might compete with each other for siting of a CSP facility and each county may not be able to handle the regulatory aspects as effectively as might be optimal from the local perspective.

5. Community Support and Opposition

Summary

The lines of support and opposition were already shaping up in the fall of 2008 when the interviews were begun. The main themes that have since emerged as the controversy has developed were fairly well described at that time. Interviewees described those supporting CSP development as county commissioners, farmers with land to sell, those working on economic development in various ways, the college, and workforce training.

Those potentially opposed were identified as the residents of the Valley because they would not benefit economically, water interests in the SLV, the SLV EcoSystem Council, residents of Crestone, the Rural Electric Cooperative, and Spanish land-grant families in Alamosa and San Luis. Early on, concerns were expressed about whether the people in the Valley would benefit economically, whether the Valley floor would be industrialized, and if the cultural heritage of the Valley would be preserved. The public meeting processes seemed to involve various organizations telling SLV residents about CSP technology and about proposed transmission line siting, with each member of the public having a turn to share his or her views. None of the interviewees stated that the public meetings processes resulted in their feeling "heard," that their views had changed, or that any of the input at meetings resulted in changes to decisions made.

5.1. Introduction

The stakeholders were queried on their views about whether the community supported or opposed the idea of CSP development. In addition, the interviewees were asked who the important stakeholders are whose views should be taken into account in CSP and transmission siting decisions. Finally, interviewees were asked to express their opinions about who favored CSP development and who, if any, opposed it.

The importance of understanding community support and opposition was underscored by one of the outside-Valley interviewees who said:

> No one is really bird-dogging this issue, because—think about it from the developer's standpoint. They want to put in as much time as is needed to be successful in a competitive solicitation and so it is unlikely that they are altruistic enough to go on a San Luis Valley-wide educational campaign without having been the recipient of an award. Once that happens, I think that if the awards are strategic, they could do just that to educate the citizens on what their technology is, what they are intending to do, but the dynamic that has happened—it takes time and it takes money, and it is not necessarily needed in order to respond to an RFP.
> Until someone wins an RFP and proposes to develop a facility in the SLV, we won't know what the people think.

In this section of the report, the following materials are covered:

- A comprehensive list of stakeholder groups
- Self protection in the Valley
- Stakeholders' characterization of sentiment toward CSP development and transmission in the Valley.

Inevitably, interviewees mentioned the low levels of public knowledge about CSP, and also shared concerns about the self-protective nature of Valley residents, and the issues that are shaping opinion about CSP in the Valley.

5.2. Stakeholders Identified Relative to CSP in the SLV

The interviewees were asked: What parties, organizations, or groups do you consider important stakeholders with respect to CSP development in the San Luis Valley?

The responses to this question have been content-analyzed. At least 125 stakeholder groups (and a few key individuals) were identified, and these represent the most important stakeholders relative to the proposed project; however, they should be viewed only as a partial list.

Although the stakeholders identified vary in the influence they might have on the CSP and transmission siting decisions, the length of the list illustrates the complexity of the web of socio-economic and environmental interests in the SLV.

Agriculturists

- Farmers in general (and their organizations)
- Potato growers
- Alligator farmer
- Ranchers (and their organizations)
- SLV Cattlemen's Association
- Colorado Farm Bureau
- Colorado Potato Administrative Committee
- 25 x '25

CSP developers/trade associations

- SkyFuels
- Abengoa Solar
- Ausra
- Interwest Energy Alliance
- Solar Energy Industries Association

Environmental organizations

- Audubon Society
- Clean Energy Action
- Colorado Natural Heritage and Rocky Mountain Bird Observatory
- Colorado Wild
- Colorado Wildlife Federation

- Crestone Sustainability Initiative
- Cultural Heritage Center
- Quiet Use Coalition
- Nature Conservancy
- National Wildlife Federation
- Playa Lakes Joint Ventures
- Rio Grande Headwaters Land Trust
- Sangre de Cristo National Heritage Group
- Sierra Club
- SLV Ecosystem Council
- SLV Citizen's Alliance
- Western Resource Advocates
- Wilderness Society
- Wildlife groups

Economic development

- Alamosa Convention and Visitors Bureau
- Bankers
- Chambers of Commerce
- Colorado Renewables Conservation
- Connecting Colorado Renewable Energy Resources
- SLV Resource Development Group
- Colorado Energy Forum (representing private investment companies)
- Optisolar
- Upper Rio Grande Economic Development Commission
- *Valley Courier* (Alamosa newspaper)

Educational institutions

- Adams State College
- Trinidad State Junior College

Elected officials

- Western Governors' Association
- Then-US Senator Ken Salazar (now Secretary of the Interior)
- US Representative John Salazar and his staff
- State Senator Gail Schwartz
- Governor Bill Ritter
- Boulder City Council, Ken Wilson (named as having an interest)
- Ex-State Senator Gigi Dennis

Federal Government Electricity Agencies

- Federal Energy Regulatory Commission (FERC)
- North American Electric Reliability Corporation (NERC)
- Western Electric Coordinating Council (WECC)

Government/Federal

- Baca National Wildlife Refuge
- Bureau of Land Management
- Environmental Protection Agency
- Federal Energy Regulatory Commission
- National Park Service
- National Renewable Energy Laboratory
- Sand Dunes National Park and Preserve
- US Forest Service
- US Department of Agriculture
- SLV Resource Conservation and Development initiative, US Department of Agriculture

Government/State

- Governor's Energy Office (GEO)
- State legislators
- Clean Energy Development Authority
- Colorado State Land Board
- Colorado Division of Labor Workforce Center
- Colorado Department of Natural Resources
- Colorado Department of Transportation
- Colorado Department of the Treasury
- Colorado Public Utilities Commission
- Department of Water Resources
- Colorado Division of Wildlife

Government/NGOs/Local and Regional

- Alamosa City Manager
- Mayor of Crestone
- County commissioners (Alamosa, Conejos, Costilla, Mineral, Rio Grande, Saguache)
- Huerfano County Commissioners
- Alamosa City Manager
- SLV GIS Authority
- Historic Advisory Preservation Committee for Alamosa
- Rio Grande Water Conservancy District
- Upper Rio Grande Economic Development Commission

Landowners

- Crestone Property Owners Association
- Trinchera Ranch (Louis Bacon)
- Billy Joe "Red" McCombs

Law firms

- Energy Minerals Law Center

- Water law teams
- Front Range attorneys

Private companies

- A&J Solar
- Black and Veatch
- Blake Jones, Namaste Solar Electric (a PV entrepreneur)
- Black Hills, LLC
- Bill Clark Trucking
- Natural Power
- SunEdison

Service organizations

- Alamosa Wastewater Treatment Plant
- Hospital in Alamosa

The public

- Citizens
- The entire population
- Electricity customers
- Local residents
- People who live near the transmission lines

Solar and sustainable resource advocates

- Southwest Energy Efficiency Project (SWEEP)
- Clean Energy Action
- Colorado Solar Energy Industries Association (COSEIA)
- Colorado Renewable Energy Society (CRES)
- Solar Alliance
- SWAT (a Southwestern transmission organization) •
- San Luis Valley Solar Association
- San Diegans for Smart Energy

Spiritual groups

- Shuma Institute

Utilities and transmission line owners

- San Luis Valley Power Authority
- San Luis Valley Rural Electric Cooperative
- Tri-State Electric Generation and Transmission Association
- Xcel Energy
- Public Service Company
- Colorado Rural Electric Association
- Colorado Independent Energy Association

Water interests

- Rio Grande Water Conservancy District
- San Luis Valley Water Group
- Water law teams of Front Range legal firms
- SLV Water Protection Coalition
- Water Watch Alliance

5.3. Self Protection in the Valley

One theme that emerged from the interviews was self-protection in the Valley. Seven of the 15 Valley respondents talked about the strong community aversion to outsiders coming into the Valley and taking advantage of its residents in any way.

Even the interviewing for this study was seen by a few Valley stakeholders as an intrusion of outsiders into local affairs. One outside stakeholder with connections in the Valley put it this way:

> Most rural communities have fear; they're skeptical of folks from outside coming in and doing stuff. It's an interesting mix of emotions and you don't want someone from the outside coming in, but you don't believe you have the capacity to do it yourself. That's true for many rural communities—it's a rural psychological mindset.

A Valley stakeholder put it succinctly:

> The biggest issue on the table is will we do it ourselves or will it be done to us and for us?

Other respondents from the Valley said the following on this point.

> We were looking at concentrated solar and what they were able to produce now in terms of megawatts and, obviously, a lot of this information came out of California, and then how much land base that was used and then multiplied that land base to come up with 5.6 gigawatts.
>
> So we started looking at that and thought—my God, that's an industrialization of the Valley floor.
>
> And of course they need water, concentrated solar needs a fairly substantial amount of water, so the valley has the three elements they are looking for which is one of the reasons that European companies[39] are starting to court the Colorado State legislature. And that's happening now. It has been happening.
>
> Our people are very vocal...if they came in and worked with us, there wouldn't be much opposition. But if an outside group tried to come in and strong-arm us, this is a tough community.
>
> We want to protect the environment and to protect against big government and corporations. If it is done properly and through the right channels...so, really, it is how it is done.

The reference to California is an important indicator that local residents opposed to a proposed transmission line to export renewable electricity from the SLV were in communication with an organized group opposing transmission lines for renewable electricity in California (San Diegans for Smart Energy Solutions).[10] Local area organized opposition groups can and do share information and resources with each other in other parts of the country. Interestingly, the issues posted on the San Diego group's web site track closely with some of the claimed advantages of the proposed transmission line into the SLV, including the following.

- Serious "power gap" without the transmission line
- Proposed line is needed to transport renewable electricity to meet RPS goals
- Protects against fire hazard
- Cost-effective and beneficial to ratepayers
- No significant environmental impacts
- Ensures energy reliability
- Good for the local economy and fosters job creation.

The opponents to transmission attempted to counter each of those points.

This preliminary evidence suggests that these themes will likely recur in areas of the Southwest wherever transmission lines for renewable electricity are proposed. Given the commitments of the Western Governors' Association and other key organizations to developing transmission to bring renewable electricity to cities, understanding the dynamics of these opposition groups is evolving as an important national issue.

5.4. Community Positions for and against Renewable Energy in General

Stakeholders were asked: "Thinking about the SLV community as a whole, and also about different groups within the community, what would you estimate is the level of support for renewable energy development in general—positive, negative, or don't have an opinion?

None of the stakeholders within or outside the Valley said that the community was opposed to the development of renewable energy in general. In fact, the general sense from the interviews was that groups in the Valley strongly support renewable energy. An illustrative quote is as follows.

> In the Valley, I would say [support for renewable energy] is extremely high. I would say a supermajority, how's that? More than a majority.
>
> Tremendous, very high—without exception folks are buying into renewable energy development.

5.5. Community Positions for and against CSP Development

Stakeholders were also asked: "How about levels of support for the proposed CSP facility?" Responses differed from the broad support mentioned for renewable energy generally. Of the 15 Valley stakeholders, only two said that the community favors CSP development, and one of those said the community didn't understand what CSP is. Support for solar development is based largely on the favorability toward the SunEdison PV plant near Mosca.

> Support for Sun Edison was high, so why would people not jump on board and be supportive of a CSP project?

A Valley stakeholder said, "The reason the Tri-state people won't build PV systems is the cost ($0.22/kWh)."

Speaking about one of the citizens at a public meeting, this stakeholder said:

> [The opponent] thinks we can run the entire Valley on solar and not have any need for transmission and most of these people have the idea that we could become isolationists and serve our own needs. Well, that's taking some steps back in time that nobody I'm familiar with is willing to take.

On the other hand, half of the stakeholders *outside* the Valley (5 of 10) said they thought the community was favorable to CSP development.

> I would say [the level of support] is high. Just from what I've heard from economic developers, from citizens, from residents. I can't say I've talked to a lot and, of course, most who have talked to me talked me up, so it's not...I wouldn't say I have rubbed shoulders with a representative sampling of the valley residents, but from what I do hear and read the support is quite high.
>
> Another outside stakeholder said:
>
> I would say mostly positive. Again, I am basing this [on my experience]. I take it that their elected representatives were in support. [At] a town meeting in Alamosa last year [there were] about 65 people...and I would say that except for the reservations....Well, there were two things. One were some reservations about transmission lines, although the crowd, on average, overall decided it would probably be worth it, and [there were] a lot of advocates there for more decentralized solar instead of just centralized stations.

A third said that the people are positive "but not terribly knowledgeable."

Nearly half (n=7) of the Valley interviewees said they believe the SLV community was mixed in its reaction to CSP development. The situation is complex.

It was said that the county commissioners need help in understanding capacity.

It was also noted that Representative Salazar's aides were hearing from opponents. Some quotes from stakeholders inside the Valley exhibiting the community's mixed reaction are as follows:

> I think there are certainly cultural issues and issues of sustainability. Commodifiying of natural resources seems to be a big concern—the commodification of solar power.
>
> *Interviewer:* Does that translate to opposition to exporting power out of the Valley?
> *Stakeholder:* Yeah, I think people are concerned about becoming an exporter of power. Again, the Valley wants to receive something in return.

Interviewer: And what does 'cultural issues' mean?

Stakeholder: Philosophy that solar energy and wind power are appropriate for farm and residential areas and sustainability, but on the larger scale it becomes more utility, becomes industrial. Even the idea of a major corporation being in control of a local resource and making money from it [is a problem]. There are some that want community control of it, some who don't care, and some who are just concerned about the scale of the development.

Another Valley interviewee expressed it this way:

Interviewer: Would the SLV community oppose a project that large?

Stakeholder: It depends on where it is located, but finding that amount of space would be very difficult. There is a large environmental contingency that is very active and would have concerns.

Interviewer: What types of concerns do you think they would have?

Stakeholder: So far they are concerned about it being an unsustainable renewable energy development, too large of a scale--large plots of land for basically one industrial purpose. They have concerns about the water as well—the quantity of and types of solutions or lubricants used. Some people are overwhelmed by the size and taking agricultural land out of production and decline of a way of life.

Interviewer: So a concern about the cultural heritage of the Valley?

Stakeholder: Absolutely, absolutely. And, if they end up putting a large facility on the side of a mountain and a big shiny thing, that would raise a lot of concern.

Interviewer: So aesthetic concerns?

Stakeholder: Absolutely, the aesthetics.

Interviewer: What about locating a facility on retired agriculture land? How would that be received?

Stakeholder: I think farmers in particular would be open to that. It's just a matter of logistics.

Another Valley stakeholder spoke of mixed reactions:

Interviewer: So would you say in general people are favorable toward CSP or not?

Stakeholder: I think there are a lot of exaggerations, a lot of anxiety, and there is a lot of misinformation...the question is, one, are we going to see generation that is simply going to be shipped out of this valley with no revenues/resources left behind...what they're worried about is can there be a way to share in the profits/benefits of the industry, long-term jobs, not just short-term jobs, and is there a way to also have revenues that will be generated and stay in the region? I have some ideas about that.... And there's a push to have more, what they call, distributed generation, to put solar panels on all the houses and the schools and the buildings and, therefore, maybe don't use utility scale. You hear that push back about we don't want to change the culture.... I think you're in a place that could go one way or the other. You could have some development here...here isn't enough...I think the *utilities are certainly going to control it because of the power purchase agreements, but the question is you want to have it integrate into the value and the culture of the region.*

Another Valley stakeholder said about the opponents:

They want it to just be for us only and not transmit anything out. They all want clean energy but they don't want to give up...have any environmental impact.

Outside the Valley the view of community support and opposition is different, as would be expected. An outside interviewee said:

I imagine it is like anywhere else. There're going to be people who are for it and people against it.

Another outside stakeholder said:

... that the citizens in the valley are highly interested and whenever there has been an event, an educational event, the rooms have been packed. They certainly understand the gravity of the situation and the need for energy and the need to educate themselves. I think they are understandably tepid about what the impacts might be, specifically to their local communities, for the reasons that I mentioned about a not well-understood technology and wanting to make sure that their local communities benefit from any development in the San Luis Valley.

Interviewer: So it is clearly understood by the people but not by the.... You mean by the people themselves.... You don't mean by the technologists?
Stakeholder:
I think it is a new technology to most people and I think there is a good amount of education or a lot of education that would probably need to happen in the valley and I think we need to turn that responsibility back over to potential developers to educate the citizens in the valley on what the technology is, what it isn't, to debunk some myths, but probably, most importantly, not to overpromise. I think...the feeling in the valley is a strong interest.

Another stakeholder outside the Valley said:

From the different kinds of groups, I'd say anybody in the economic development area would probably be supportive with some caveats probably, but...

Interviewer: What kind of caveats?

Stakeholder: I think that what I hear is probably a realistic concern, is that when there is a large plant to be built, of any kind, you can bring in labor, and can tend to come into an area, overwhelm the area (maybe, maybe not) and then leave the area. So, there's a kind of preference for local labor and that doesn't always happen. That was the case in Nevada for Nevada solar power, so that would be one of those caveats.... I think you might find groups that are leery of companies coming in and taking advantage of the area."

5.6. Don't Know What Stakeholders Feel

Three of the interviewees (two outside and one Valley) said that they didn't know what community response toward CSP development was. An outside Valley stakeholder said:

I don't know because I haven't wandered around there and asked people. At [other sites] public support has been tremendous, very positive. Economic growth, tourism, clean energy, it has been exciting. I think the same will be true in the Valley. Whoever develops the plant has to start meeting early with the local people. It takes a long time to explain to people—[questions arise], such as will the mirrors blind pilots? [One developer] had to fly over [a] plant and take a video showing that at no angle would the sun's rays be reflected up to the airplane.

5.7. Perceived Need for Education

Several interviewees stressed the need for public education. For example, one of the outside stakeholders said:

There might be a misperception by the public that these renewables are an answer. I don't think they understand the impact that, especially solar, has on the land. And that is you know—one square mile of solar facility equals so many megawatts and when you really look at that it's a square mile of nothing but a solar facility. There's not too much else that happens on that land.

Two Valley stakeholders said:

They don't know enough. If people were honest with you, I think they would tell you that. They just don't know because of lack of education. Adams State College could play a role with that if they wanted to—could be a place where educational forums could take place.
...there is no group that is standing out there saying we need concentrated solar power. There is no group that even probably knows what it is.

Pointing to the need for public education, another interviewee from the Valley stated:

[There was] a public meeting last night.... Some of those speakers from Crestone there...had no desire for any kind of transmission project and had the desire that we produce all of our energy from our own source.... [There have been] more and more meetings. Last week [there were] two meetings on transmission lines and all these issues get addressed every time [there is] a meeting.... [there will be] another series of meetings basically on the transmission line, [I think there should be] a focus group of people to deal with energy issues because there are huge issues today.

6. Transmission Issues Raised by Stakeholders

Summary

The main reason for CSP development in the Valley is the provision of electricity to Colorado's Front Range cities.

Stakeholders outside the Valley accepted this premise. Within the Valley, support for more transmission lines into the Valley revolved around the desire to increase the flow of income into the community by exporting electricity, the need for increased security and redundancy of the Valley's power supply, and the need to improve the Valley's grid and increase its energy supplies. Opposition to transmission lines focused on fear of loss of control of local land and of the decision processes, aesthetic concerns, fear of industrialization of the Valley floor, wildlife concerns and concerns about the legal processes at the Colorado Public Utility Commission regarding the proposed large-scale transmission line over La Veta Pass and into the Valley from the east.

6.1. Introduction

The interviewees were queried on their views of the most important considerations in the siting of a transmission line for renewable electricity. They were asked: "From the perspective of your organization, what are the most important considerations involved in power transmission?" Responses ranged from full support for the transmission through neutral, factual responses, to mention of community conflict about transmission, to opposition.

Stakeholders in the Valley said that Rio Grande, Alamosa, and Saguache Counties are interested in CSP development. It was generally agreed by knowledgeable stakeholders that new transmission lines would be needed if a CSP facility at a scale of about 200 MW were to be developed. One said: "The current Climate Action Plan models a 200-MW concentrated solar power facility somewhere in Colorado."

Viewsheds, availability of water, and availability of transmission are the biggest considerations for siting CSP. This section discusses the issues raised in the interviews relative to transmission on a continuum from support to opposition.

6.2. Support for Transmission

As reasons to support transmission, Valley interviewees perceived the following: (1) the desire to export power as a way to bring income into the community, (2) the need for security and redundancy of the Valley's power supply, and (3) the need for improved power infrastructure.

6.2.1. Desire to Export Power

We located 20 comments from stakeholders inside the Valley that addressed the need for transmission to export power. This is one of the most commented-on issues in the interviews. One of the Valley interviewees said:

> Well, in some ways we have the ability to meet our needs with solar and provide for other people's needs, so we have done a little bit of research, even with people that live in places like Boulder and asked them if I have a megawatt of power and I'm generating in the San Luis Valley, would you have an interest in it? And we've got responses that include things like not only would I have an interest in it, if it is really renewable and can really guarantee that this project could be sited in the San Luis Valley, I will pay you a premium. I will give you a $1.30 for your $1 worth of power and I will buy it by subscription, I will pay you a year in advance so you have the money to help build the facility.Yeah, it's about 200 MW as a rule of thumb for a number to use. So if you went to 200 MW, doubled its size, you would meet the demand and then you would have the ability to expand to 400 megawatts and instead of importing 200 you'd be exporting the extra 200 so now your renewable energy generator is exporting 200 MW and generating the same income that used to leave here as a payment to a utility.
>
> Interviewer: Okay, so the idea is to own [one or more CSP facilities] locally and first meet local demand.
>
> Stakeholder: That is going to be the most palatable environmentally.
>
> Interviewer: Okay. What about transmission? I mean transmission is a big issue or apparently a big issue, so what about that?
>
> Stakeholder: Well, 100 MW, should we put this on the ground, we meet all the conditions that you and I described to date, then we would be meeting half the demand of the San Luis Valley.... And second export and bring money into the Valley by selling power.

An outside stakeholder described it this way:

> Really no one can build anything in the SLV until there's a transmission highway to get it out.

6.2.2. Need for Security and Redundancy

Valley stakeholders talked about their concerns about the vulnerability of the Valley's electricity supply. Ten comments fell into this category. One interviewee from the Valley said:

> *Interviewer:* You mentioned transmission as a limiting factor?
>
> *Stakeholder:* Yeah. But there is a proposal for Xcel and the REC to go together and put a pair of 230-kV [transmission] lines over La Veta Pass from the east, which would be a

good idea because of the vulnerability that we are in right now because it all comes over from the north.

Another Valley stakeholder said:

It comes from the north, two transmission lines, and last night we even left the Homeland Security comment on the table that two guys with one chainsaw could take all the power out in this valley. There is no redundant power. All the poles....if you cut four telephone poles you can shut down the power in the entire San Luis Valley. It is completely inappropriate for any future development. There is a huge significant issue about redundancy in transmission lines.

Another stakeholder was asked if they had any thoughts or feelings about the energy leaving the SLV to help Denver or the Front Range. The response was:

Yes, positive as long as it promotes revenue. But concerns about transmission—what will it take to handle the transmission of solar power. We don't want to export the water. If there is a win-win then the Valley would be all over it.

Another Valley stakeholder said:

So the Valley basically right now uses about 155 MW and I believe you need 1,000 MW to equal one gigawatt, so then the question becomes, geez, what is all that other potential for? Well part of it is redundancy so the Valley has another way to get energy if something happens; the other is for the Valley to grow and possibly invite industries that have never been able to locate here before. That's what's being promoted. And then the other piece is that it is a valuable potential to be able to export power. For the first time the Valley will become an energy producer, and obviously one of those ways the Valley will be able to produce energy is through concentrated solar.

6.2.3. Need for Improved Power Infrastructure

At least one Valley respondent said that the SLV needs more power.

Interviewer: So they must be thinking that there is going to be renewable energy development in the Valley if they're building this line.

Stakeholder: We needed the power in as much as eight or nine years ago, so the initial thrust for that line was us needing power to support our developments, our growth in the western area, South Fork, Creede. When renewable energy came around and NREL named us a spot for really good solar and they passed the renewable energy bill, all of those things said to Xcel we're going to need transmission from the San Luis Valley.... [It will happen] one way or another. And it should be.... we don't put any money in, it is just for our system, and Xcel and Tri-State pay for it. It is about an $80 million project. It should be completed by 2012 or 2013.

Interviewer: And why are they doing this?

Stakeholder: Because we are running out of power in the valley. The last transmission line built in San Luis Valley was 35 years ago.

Although existing transmission lines could probably support an additional 50-100 MW of CSP development, Valley stakeholders made several comments suggesting that the size of the CSP plant needed to be as large as 200 MW.As noted earlier, this size of facility would allow local demand to be met first and would begin to yield additional electricity for export out of the valley.

Another local stakeholder talked about the scale of a possible CSP facility.

Is this the one [a Valley expert] keeps talking about that is 5,000 acres and enough power to power six planets? ...We talk about it all the time because it is considered very important for our economic development. If you take a resource like that and don't leverage it, then you're really missing the boat. We are "green" here in the Valley and it resonates with us. Our agriculture industry is having problems because of water and depletion of the aquifer. And since agriculture is our number one industry, we need to diversify.

Tourism is our number two industry, which only represents 11% of our economy. We have 80,000 square miles here in agriculture production, so...if some of these farms converted to a product that did not use so much water that is a good thing.

An outside stakeholder said:

What sets solar thermal apart from PV? PV uses semiconductor—electrons — generates electricity directly. Solar thermal is a large-scale parabolic trough—above 5 MW it's cheaper than PV. Below 5 MW, PV is cheaper. So solar thermal gives you the scale. 200 MW is the sweet spot. CSP has scalability. You would never have a 200 MW PV system.... The characteristics of thermal are that it has more thermal inertia—in the tube is oil at 750 degrees. If a cloud goes over it, it still stays hot, so it has more even performance. PV production is spiky because when it is cloudy, the output drops dramatically. So each time a cloud passes over a PV system, the production goes way down.

Thus, the grid has to be able to handle spikes. There is no such limitation for solar thermal, it's very predictable and even....Stakeholders said that, in the Valley, the transmission issue was as big as the water issue. Yet, it seemed to be fairly generally agreed that new transmission had to be built if a CSP plant of any economic size (e.g., 200 to 300 MW) is to be sited in the Valley. Looked at another way, a CSP plant up to 200 MW could be accommodated on existing lines, but anything beyond that could not.

6.2.4. Neutral Comments

Two quotes below illustrate a neutral, fact-based explanation of the transmission situation without exhibiting a position.

It's been all over the newspapers that the SkyFuel wants to do 1 gigawatt power— they won't do it all at once though because we don't have the transmission to handle that. They are going to have to build out if that ever happens. But there again, that's 5,000 acres.

And,

> Number one is current availability of transmission, and we are currently maxed out. There are three coming over Poncha Pass—small, medium and larger. There is one proposed to come over La Veta Pass.... Second concern is size of a new transmission line—can it meet current demand and possible new development. I think solar development will go beyond the capacity of the new line (230 KV) that Tri-State and Xcel are proposing. If at some point we are going to develop solar power and transport it out of the Valley, we need to address this issue. But how do we finance this?

6.3. Opposition to Transmission

As reasons to oppose transmission, Valley interviewees mentioned:

(1) fear of industrialization of the Valley floor,
(2) aesthetic concerns,
(3) feared loss of control,
(4) wildlife concerns, and
(5) legal-regulatory concerns.

6.3.1. Fear of Industrialization of the Valley Floor

A few comments appeared to oppose transmission lines. Fear of uncontrolled industrialization of the Valley floor was expressed in the following quote:

> I don't think anybody in the valley has a problem with a one gigawatt transmission line. The problem is once that transmission line gets in here, what are the possibilities of taking advantage of that because the loop will then all be there, the loop will be formed, and then other industries will be very attracted to the fact that it is in now and they may want to develop and build on that and they may want to take it to 5.6 gigawatts.... All these companies aren't going to invest in the valley for 1 gigawatt. They're not. They're going to want more and that's the quandary. And unfortunately that's what concentrated solar represents.... So are we really thinking in an antiquated way by thinking in terms of concentrated solar? Will we end up coming up with much more efficient ways of being able to distribute solar other than by concentrating it?

Also, concerns about public health and electro-magnetic fields (EMFs) were said to be concerns of neighbors living near the proposed lines.

6.3.2. Aesthetic Concerns

A few comments decried a proposed transmission line because it would be "on our vistas" and would be aesthetically displeasing, "going over the mountains."

It was noted that one of the Valley's counties opposed transmission lines.

> Costilla is most concerned with transmission lines being placed across their county—aesthetic concerns, and also they want to be sure that they receive local benefits. In particular, their largest landowner, Mr. Bacon is concerned about transmission lines

going across his private land, which is used for ranching and wildlife. He might have concerns about future property development.

6.3.3. Fear of Loss of Control

As noted earlier, Xcel Energy and Tri-State would be granted eminent domain should they be granted the CPCN for the transmission line. Some comments expressed fear of loss of local control of property legally condemned under eminent domain.

6.3.4. Wildlife Concerns

Concerns were expressed about the possibility that the transmission line would bisect wildlife corridors in the La Veta Pass and Trinchera Ranch areas and that transmission lines can be hazardous to birds. According to some residents, power poles can give raptors an advantage over the other species because transmission lines are required to provide raptor perches.

6.3.5. Legal-Regulatory Concerns

As described in Part Two of the report on the transmission controversy, Louis Bacon, who owns the Trinchera Ranch, has hired attorneys to oppose Xcel Energy and Tri-State's application for a CPCN to build the transmission line. In interviews, Valley stakeholders had identified Louis Bacon's opposition to a proposed transmission line.

7. CSP Facility and Transmission Siting Decision Models

Summary

One of the key points from the interviews is that the facility siting and transmission siting decision processes are extraordinarily complex and are not well understood by the public. In fact, those responsible for decision-making may not themselves fully understand them.

They may be opaque to the public, which makes public comment more difficult. Only a few parts of the decision processes appear to be open to public comment. Knowledge of how these decision processes work equates to power for those organizations "in the know"; local community groups may not have access to those processes in a timely way.

7.1. Introduction

During the interviews, stakeholders discussed how they believed the decision to site a CSP facility would be made. At least three different decision models emerged from the interview data: (1) SLV Power Authority decision-making, (2) county permitting, and (3) permitting on public lands. Each of these is described in this section. In addition, other organizations have decision authority affecting the transmission and siting of a facility.

7.2. SLV Power Authority Decision-Making

A number of stakeholders within and outside the Valley mentioned the concept of a locally owned 100-MW CSP power plant that would sell electricity. A concern about the security of power supplies in the Valley was one motivation for moving ahead. As one community member put it: "Any time you can shut an entire population of 18 incorporated communities with a chainsaw, cutting four poles.... and worse than that...somebody rolling a car and starting a grass fire that could develop into a forest fire would take it out also."

The only transmission line into and out of the Valley goes over Poncha Pass. As noted earlier, Valley stakeholders mentioned that all the Valley's power comes in via the existing transmission lines from the north, which leaves the power supply—which lacks redundancy—vulnerable to sabotage, forest fire, or other unexpected occurrences.

Under existing statutes, a local nonprofit was working on forming an SLV Power Authority.

The primary statute relied upon authorizes a power authority that ties together the cities and counties for the sole purpose of providing a "nest." As the leadership becomes more aware of the potential of developing solar power for their own purposes, the Power Authority would be available to them.

> So going again back to 1995 when we first formed the authority and statute, so the county commissioners could, for the purpose of telecommunications or power, have a body that could manage an authority. It actually became public on March 1, 2006, when the National Renewable Energy Lab and Tri-State Generation and Transmission, Rural Electric, Xcel Energy, Healthcare Governance and Education all came together at Adams State College to talk about the feasibility of doing a 100-MW concentrated solar project that would be a locally owned, locally controlled project that would redistribute power to the existing utilities—the Rural Electric Cooperative and Xcel Energy—as well as some of the people off the grid. Historically, we got a certificate of area where some of the primary concentrated populations were served by Xcel Energy and the rest of the valley was served by San Luis Valley Rural Electric Cooperative.

At the time of the interviews, during the fall and winter of 2008-2009, the SLV Power Authority was still being developed, although most of the 18 county commissioners had put forward the money to start an RFP for a feasibility study.

A local concern was, reportedly, that SkyFuels and Ausra (CSP developers) had been in the Valley discussing the possibility of a 1-GW CSP facility.

> The biggest issue on the table was will we do it ourselves or will it be done to us and for us. We actually formed a council of governments on July 28, 2008, to formalize the revised statutes to enable the county commissioners to form the power authority.

The concept that the local stakeholders were working on involved siting small 0.2-MW projects at all 15 substations in the Valley—called "diffused solar"—to make more than enough power to meet the valley's needs. A local stakeholder said:

> ...amazingly people here control hundreds of millions of dollars in property or hundreds of millions of dollars in finances, but when it comes time to do a $100 million project will balk that they don't have that history...

An institutional barrier to "diffused solar" was described by an outside stakeholder. He said that the rural electric cooperatives are sold electricity generated from coal plants built with loans from the Rural Utility Services (RUS), an agency of the USDA. As part of RUS funding coal plants with low- or no-interest loans, the RUS institutes an "all-requirements" contract through which cooperative members agree to buy 95% of power for the life of the project. RUS funds Tri-State, which distributes electricity to 44 rural electric cooperatives. In this way, the 44 co-ops are, in this stakeholder's view, "yoked to coal plants." Although the co-ops could bring on 5% of power from another source, this stakeholder said that the regulations to do this are so onerous that the co-ops cannot do it. This stakeholder believed that exemptions from the "all-requirement" contracts are needed at the federal level for cooperatives to put "green electrons" into their systems.

A related concept is that of a distributed system in which, say, 20-MW plants could be built next to existing power substations. In this more distributed model, the dollars generated would stay in the state's economies and would be spread across the tax base of the local counties, which would keep funds within the state. There could be a higher cost for wholesale electricity, but the benefits would far exceed those of the centralized model where so many of the dollars go out of the state.

Another outside stakeholder said, in reference to the idea of a locally funded CSP project, "trying to raise $1 billion is very challenging." A CSP development project is very complex, involving independent engineers, private auditors to estimate market value, and "it will cost $1 million in legal fees alone."

Another outside stakeholder raised a question about whether a local power authority could actually own a CSP plant.

> Well, we've got San Luis Valley Electric Co-op serving part of the valley. We've got Public Service Company serving part of it, currently Alamosa, and if you are talking about somebody who would serve their territory, you're going to have to get legislative[41] approval to do that. They are not a utility now and in order to become one, if that's their intent...There are lots of ways of reorganizing entities like cities into municipal utilities. It means negotiating with the utility that today holds the franchise and condemning the property, all of that stuff, but if they want to be just a wholesale supplier to the grid, they've got to find a utility to buy their power. If they want to become a utility, as in a distribution utility, there's not enough people in the San Luis Valley to use that much power so it's got to go out of the valley somehow, doesn't it?

This model of locally owned and operated small-scale CSP facilities distributed over the Valley floor was a popular concept among Valley stakeholders.

The situation that was described by stakeholders outside the Valley, however, was that the Valley electricity customers could not absorb all of the electricity from a large-scale CSP facility; therefore, a new transmission line would be needed.

With respect to development of a local authority, existing statutes may give some legitimacy to this effort. Renewable electricity cooperatives have been authorized by

Colorado statute to invest in generation assets and sell power at wholesale. This would be regulated by the *US Federal Energy Regulatory Commission (FERC)* at the national level; FERC regulates the wholesale power market.[42] FERC is an independent agency that regulates the interstate transmission of natural gas, oil, and electricity. FERC also regulates natural gas and hydropower projects. Chairman Jon Wellinghoff recently issued a statement about the integration of renewable resources into the nation's electricity grid:

> Approximately 9,000 MW of variable energy resources [VERs] went into commercial operation in the US during 2009 and a similar number in 2008. Currently, there are thousands of MW of VERs, particularly wind, in the process of interconnecting to the electric grid. These new resources offer low operating costs and low greenhouse gas emissions, which can benefit consumers.... VERs also have some operational characteristics, which present challenges to system operators. Therefore, it is important that the Commission examine the most efficient ways to effectively integrate these resources into the electric grid, while maintaining reliability and operational stability. We want to examine whether existing rules, regulations, tariffs or practices within the Commission's jurisdiction hinder such efficient integration. I look forward to parties' comments on the issues and questions that the NOI [notice of intent] identifies, and to thoughtful and thorough discussion of any reforms that may be needed to remove barriers to the full integration of VERS.

The *North American Electric Reliability Corporation (NERC)* manages the standards for the system.[43] It ensures the reliability of the bulk power system in North America. NERC develops and enforces standards, assesses the adequacy of the electricity supply, monitors the bulk power system, and educates industry personnel. It is subject to oversight by FERC, which granted NERC the legal authority to enforce reliability standards of all US users, owners, and operators of the bulk power system, making standards mandatory and enforceable. It is a non-government organization with statutory responsibility.

The *Western Electricity Coordinating Council (WECC)* markets hydropower.[44] WECC is the regional entity responsible for coordinating and promoting bulk electric system reliability in the Western Interconnection, a synchronized electric grid that spans 1.8 million square miles from the Canadian provinces of British Columbia and Alberta, to the northern part of Baja California, Mexico.[45] In addition, WECC assures open and nondiscriminatory transmission access among members, provides a forum for resolving transmission access disputes, and provides an environment for coordinating the operating and planning activities of its members as set forth in the WECC Bylaws. Membership in WECC is open to all entities with an interest in the operation of the bulk electric system in the Western Interconnection. All meetings are open and anyone may participate in WECC's standards development process.

To understand these complex relationships, the independent power producers (IPPs), the investor-owned utilities (IOUs), and WECC all compete to sell power on the electricity markets.

A San Luis Valley Power Authority would be an IPP competing to sell power. Although there are Colorado statutes permitting municipalities to form utility districts, it is less clear that counties or multi-county regions have the statutory authority to do so under Colorado law.

7.3. County Permitting

A different model of CSP development is that a private CSP development company (IPP) purchases sufficient privately owned land, along with water rights, on which to site a facility.

This subsection describes interview findings on the role of CSP development, Xcel Energy's role, and the role of the Colorado Public Utilities Commission (PUC). A local stakeholder, who seemed to believe that PV and CSP were interchangeable, said that numerous companies (IPPs) had been acquiring private land.

> We've had 10-15 companies who have approached us, been in to talk to us, and acquired land throughout the valley [to build solar collectors] I think already or put an option on it....

In this model, the usual permitting processes controlling any facility siting application is used for the CSP plant. One Valley stakeholder described it in the following quote.

> Well, [there is] zoning and this type of stuff and the control where different things are set. We run it before the planning commission and then it comes before [the county commissioners] and the planning commission has public meetings and hearings to address whatever concerns the residents...if somebody comes and picks out a location and this is where they were going to set it, they would have to come before the planning commission, take out the proper permits. [The county commissioners] can either deny it or approve it....

Two major comments from study stakeholders have a bearing on this process. One, from a Valley resident, said that the county commissioners tend to avoid their role in public policy.

> They just avoid everything and the way to avoid everything is just to say 'I don't want any regulations.' Unfortunately, what they end up doing is inviting all their decision makers to come in and make a decision for the rest of the county...County commissioners have a lot of authority, but...they don't exercise it on the level that they could. County commissioners
> .... As a general rule are rather disempowered in this valley. It is kind of sad. There are a couple of personalities that really step up to the plate but a lot of them are really timid. They are always afraid that they're going to be sued so they have a tendency to fall into this place of inertia.
> They don't want to be sued by industry. That was said to us when we were working on the [an environmental] issue. We actually had the attorney say, well, we'd rather be sued by the environmentalist than...the developer."

The other comment pertained to the realities of utility decision-making in the state. All of the electricity customers are already being served by the SLV Electric Cooperative and by Xcel Energy.

If someone else tried to serve the SLV territory, then there would have to be legislation to permit a new utility to exist.

> There are ways of reorganizing entities into municipal utilities. It means negotiating with the utility that today holds the franchise and condemning the property, all of that stuff,

> but if they want to be just a wholesale supplier to the grid, they've got to find a utility to buy their power.
> If they want to become a utility, as in a distribution utility, there's not enough people in the San Luis Valley to use that much power so it's got to go out of the valley somehow, doesn't it? ...There has to be a purchase power agreement if they are selling to a utility which actually has customers.
> If they want to become a utility, that means they have customers and they got those customers somehow which means somebody else has given them up...if they built a solar plant of any reasonable size, you've got to sell power outside of the valley. The valley can't absorb [that much power]."

Finally, counties do have the authority of local eminent domain, in which they can take property for public purposes.

7.3.1. The Role of Private CSP Development

An outside stakeholder described four roles in getting a CSP plant going:

1. Consultant/technologist partner (the CSP part)
2. Developer - this partner plays the role of securing land, water, transmission, permits, and the power purchase agreement (PPA); this partner could be a well-backed developer
3. A company to construct the plant
4. Owner/investor who finances the plant.

Financing a plant would require large investors. The developer pays the CSP and engineering companies, and funds the costs of getting options on the land and water permits. All of these parts would cost perhaps $2 million.

The cost of building a CSP plant is $4,500/kW, assuming a system without storage. With this estimate, the cost would be approximately $450 million for a 100-MW plant. One billion dollars would be needed for a 200-MW plant.

According to this stakeholder, only the big investors (pension funds and hedge funds) can invest in these projects.

> Companies rose up to service large plants and engineering companies got good at building them. Developers got efficient at developing these projects. They learned to borrow money, build the plant, and arrange a PPA with the utility company. They used the contract to finance the construction of the plant and pay it back from revenues from selling the electricity.
> It was a model of finance that was used over and over.... The regulated utility has a monopoly, so they can build a plant on their large balance sheet; basically it's financed through the rates, with the charge passed on to the customers.

Independent companies get a PPA contract from the utility; then go to Wall Street and get financing; then the revenues from electricity sales go to service the debt and make a profit. The same stakeholder went on to say:

> But, the business model didn't pan out. The plants were built not in the best places to sell electricity to customers. These power plants became "stranded investments." Enron was a chimera. They couldn't buy electricity from the desert and sell it in California and make a profit. One reason was the inadequacy of the transmission lines—this was a factor in the failure.

FERC also defines the "large generator integration process (LGIT)" that guides the process that leads to interconnection agreements. The interconnection agreement is between the developer and owner of the transmission line. In some cases, the utility owns the line; in other cases it is some other entity.

An outside stakeholder said that, relative to distributed technology that some people in the Valley want, "PV panels won't have a significant impact on climate change. They won't solve the problem because of the spinning reserves issue." This stakeholder went on:

> It's a billion dollar venture. I don't see local ownership. On the other hand, I guess—I'm not creative enough to think how the public could have a set tiny percent —I don't want to go there. Financing a plant in today's market is challenging. I don't think local ownership is a good idea—it's not appropriate. It doesn't fit the CSP plant—you don't have local ownership of a nuclear plant.

When asked about the effect of the current US economic situation on developing CSP in the SLV, this stakeholder said:

> It's challenging; the banks have not been lending money and when they do it's for higher interest rates and a shorter term. There is a problem on the debt side. [With respect to the 30% federal investment tax credit]...It's like getting a library card to a library that's closed. There's no way to turn it into money. The tax investors are gone, and they were Lehman Brothers and AIG; the stimulus bill has provisions being worked out; DOE has a loan guarantee and they'll be slow to work out the program. Financing is a challenge. It can be done but it's a lot different than it was a year ago. People can't borrow money for a car or house, so how can you borrow money for a power plant at $1 billion.... We don't know how the stimulus bill will work, and it's short term. Can you get a plant negotiated and get a permit by the end of 2010?...For the banks, there's a loan guarantee—but how much will they guarantee—it's still being invented....
>
> It's a challenge, these are expensive plants, there's a utility that's been told to buy green energy in Colorado, and if the financing—the fuel for the plant is money (debt and equity), and if the cost of the capital is too high, the utility will say, this is too expensive for the ratepayers. There is an issue that the stimulus bill and all the help the feds do; the feds subsidize the competition; this is a reference point; somehow there's got to be a way to finance these plants.

7.3.2. Xcel Energy's Role

A privately-owned utility, Xcel Energy is a regulated monopoly. Xcel Energy issued an RFP in January 2009 for IPPs to propose projects from all sources with bids due back in April 2009. This was driven by Amendment 37 legislation and their resource plan filed with the PUC (Docket No. 07A-447E). The utility received bids for 12,000 MW of power from all sources. Bids included fixed PV, one-axis PV, trough CSP, dish stirling, and power tower. As it pertains to the San Luis Valley, the plan is for is 280 MW and 310 MW of incremental solar resources across a variety of technologies and developers that would be producing power online by 2015. The amount of solar resources could go higher (closer to 380 to 390 MW) if the PUC directed it. The bidders are responsible for delivering the power to Xcel Energy's transmission lines, and if the site is not next to a transmission line, the bidder would have to build the infrastructure to deliver the electricity to Xcel Energy's system.

> A 200-MW solar thermal plant with storage is probably going to take about 4 square miles so it is not like every project that is bid just happens to lie right next to one of the transmission lines, so it is the capacity of the line plus where they are locating, where they are proposing to locate the plant, where the substation is going to be, how much connection is going to be required....

According to an outside stakeholder, the IPPs:

> ...have all picked their technology; they want to propose and they have picked the locations that they want to propose...as part of their due diligence, the utility will make sure that if they don't currently have site control that they have a clear path to which they can achieve site control.

A decision on the winning bidder should be made soon.

7.3.3. The Role of Colorado Public Utilities Commission (PUC)

Brief background information is needed to understand the role of the PUC in the siting of CSP projects and of transmission lines. The PUC authorizes the acquisition of 200 MW of solar with storage. The planning process employed, called Least-Cost Planning, means that literally the least-cost path to generation requirements are selected. In 2007, the PUC amended its rules to allow a deviation from least-cost which was cited as an exception to the statute. Section 40-2-123 authorizes and exhorts the commission to approve the acquisition of new, clean, renewable energy resources that could be deemed demonstration projects of technologies that were not available at commercial scale. These are called "Section 123's." It is generally acknowledged that these projects are going to be more costly than the equivalent amount and capacity of a fossil plant, but they can be selected to give new technology a start in Colorado. Bidders are asked to bid competitively for this product. According to this process, 200 to 600 MW will be coming in over the next eight years.

Any PPA that involves selling electricity to Xcel Energy would be subject to regulation by the PUC. But if the local power authority sold directly to WECC it could avoid PUC regulation.

The PUC might ultimately be the decision authority on the proposed transmission line into the Valley. According to an outside stakeholder, before the utility starts acquiring rights of way and land to build a transmission line, it needs to obtain a certificate of public convenience and necessity (CPCN) from the PUC. The utility has to get approval from the PUC for the line at a "conceptual level." The PUC considers the economics of the proposal, the reasons for building the proposal transmission line, and the impact on the grid.

If the PUC approves the line, then the utility acquires rights-of-way if need be, in a step-by-step process working through landowners, counties, and cities. Typically, the utility pays localities for use of rights-of-way, and if need be condemns land and compensates landowners. If there are disagreements, the utility has the right to bring them to the PUC to determine whether conditions attached by the localities make it economically infeasible to build the line. The PUC essentially has backstop authority over such issues. The PUC's charge is to uphold the public interest; it is not necessarily to uphold the interest of the localities *per se*.

Another stakeholder characterized the current energy situation and the role of the PUC differently. This stakeholder said of the state's utilities:

> They don't share resources. Classic example right here in Colorado: while we sit here having Xcel having 500 MW of excess capacity from Comanche 3, we have Tri-State trying to build 700 MW of coal plants in Kansas to bring the power into Colorado to serve their growing needs on the front range for the rural electric members.... Court decided against it. They're still appealing it in Kansas. Now Tri-State has announced it wants to build a nuclear plant near Lamar. We are sitting here in a state with 26 GW of undeveloped renewable power and we have Xcel that has excess capacity and Tri-State who needs capacity, and Xcel who could firm intermittent resources—if you just look at Colorado as the sandbox, you'd say "Xcel and Tri-State, you play together. You figure out how to give us the maximum amount of renewable energy at the least cost.

7.4. Permitting for Solar Energy Development on Public Lands

On April 4, 2007, the US Department of the Interior (DOI) issued an Instruction Memorandum (No. 2007-097) to its field officials concerning "Solar Energy Development Policy."

The policy background in the memorandum refers to an overall strategy "to develop a diverse portfolio of domestic energy supplies" in the United States, citing the National Energy Policy of 2001 and the Energy Policy Act of 2005 (P.L. 109-58, August 8, 2005).

Section 211 of the Energy Policy Act of 2005 "encourages the approval of at least 10,000 MW of non-hydropower renewable energy projects on the public lands within the next 10 years."

The memorandum states: "the BLM's general policy is to facilitate environmentally responsible commercial development of solar energy projects on public lands and to use solar technology on BLM facilities where feasible."

The memorandum instructs field officers to process applications for such facilities as right-of-way authorizations under Title V of the Federal Land and Policy and Management Act (FLPMA) and Title 43, Part 2804 of the Code of Federal Regulations (CFR). Applications for CSP must comply with the BLM's planning, environmental, and right-of-way applications requirements.

The instructions state that these applications "will be identified as a high priority Field Office workload and will be processed in a timely manner." They also state that "early public notification and involvement of local communities and other interests is also important in increasing public acceptance and avoiding potential conflicts, especially in areas where other uses exist on the public lands."

As of October 2008, no applications for large-scale CSP projects had been received by the BLM office in the San Luis Valley.

7.5. Permitting for Transmission on Public Lands

The USDA's Rural Utility Service (RUS) finances electricity development through the Rural Electrification Cooperatives. RUS is under the USDA Rural Development Utilities Programs. The Rural Electrification Act of 1936 provided funding through cooperative electric power companies (or co-ops), to electrify the rural areas of the United States, which before then had no power. The member-owned co-ops still exist today; their role is to purchase wholesale power and distribute it using their own network of transmission and distribution lines.

The USDA's program now helps rural utilities to expand and update their technologies, working in partnership with nonprofits, public bodies, and for-profit utilities.

Because of its role in financing Tri-State Generation and Transmission Association, Inc., which supplies power to rural electric co-ops in several states, including Colorado, and because Tri-State is one of the applicants for a CPCN to run a transmission line into the SLV from Walsenburg, the RUS was required to hold public hearings in the area of potential impact from that line.

7.6. Other State Organizations with Potential Decision Authority

Beyond the decision models described above, there appear to be other organizations with decision authorities of various kinds that will influence CSP development in the Valley. These include the Governor's Energy Office, the Water Conservation District, economic development groups, the Colorado State Land Board, the Colorado Department of Transportation, the Colorado State Treasurer's Office, the Clean Energy Development Authority, and transmission planning groups.

7.6.1. Governor's Energy Office (GEO)

The GEO does not have an impact on facility siting decisions directly. State government is not involved in a competitive solicitation (put out by Xcel Energy). However, the GEO has a program manager dedicated to renewable energy transmission issues.

Colorado's Governor Bill Ritter had made the "new energy economy" a hallmark of his election campaign, and his energy office was tasked with carrying out his goals. The state government has a vested interest in ensuring that the first CSP project in the state is sited in an appropriate manner and goes very well.

In discussing the state's role, a stakeholder said that, although GEO's role is to facilitate education on CSP, it is thought that the people in the Valley "need to hear from the developers exactly how much water, exactly what the build schedule is, exactly how tall the troughs are...." Another interviewee said that GEO could be a point of referral for SLV groups toward CSP developers to get information from them.

The public's position on a CSP facility will not be known until someone wins an RFP and proposes to develop a facility in the Valley. One stakeholder thought the GEO could become a liaison or representative of the Valley in the CSP decision process.

CSP developers were already purchasing land in the SLV during 2007 and 2008. Several stakeholders mentioned a "tremendous land acquisition" that has been occurring in the

Valley, especially prior to the issuance of the Xcel Energy RFP during January 2009. One interviewee said:

> There is a very strong political desire on the part of state government to bring in a CSP facility to the Valley.

GEO does have a financing role with respect to proof-of-concept renewable energy projects, and is the funnel of federal American Reinvestment and Recovery Act (ARRA) funds to projects within Colorado.

7.6.2. Clean Energy Development Authority

H.B. 09-1300 authorizes the Clean Energy Development Authority (CEDA), as a Colorado transmission authority. The bill was introduced to the legislature in April 2009, sponsored by Representatives Solano, McFadyen, and Rice; and Senators Schwartz, Romer, Heath, and Veiga. Transmission authorities have recently formed in neighboring states as vehicles for transporting renewable energy needed to meet state renewable portfolio standards.

Legislators in eight states have created transmission authorities since 2005, primarily in the West. H.B. 1300 would strengthen CEDA so the Authority could finance transmission projects connecting Colorado's renewable energy resources to loan centers.[46]

Public Service Company of Colorado pushed an unfriendly amendment to the bill to give the utilities the right of first refusal (RFR) before CEDA can proceed with financing projects.

If the amendment passed, the effect could signal the demise of CEDA and future opportunities to its financial support for projects that transmit clean energy. The CEDA board is unanimously opposed to the amendment because:

- Independent transmission developers seeking CEDA financing would not participate because they would not want to reveal proprietary development information to utilities.
- A utility RFR could subject CEDA to unnecessary delays.
- A utility RFR could have the effect of an exercise of undue market power control, potentially leading to an anti-competitive outcome.
- A utility RFR could lead to an exercise of undue control by a utility over a public body composed of individuals appointed by state elected officials.

HB09-1300 provides strike-below amendments to HB07-1150 that created CEDA. The CEDA Board said that CEDA should exist because:

- Transmission authorities have recently formed in neighboring states;
- Legislators in eight states have created transmission authorities since 2005, primarily in the West
- CEDA needs to finance transmission projects connecting Colorado's vast clean energy resources; Colorado should strengthen CEDA to accomplish this.
- The ARRA has opened new opportunities for CEDA.

7.6.3. Water Conservation District

The Water Conservation District has a key role in ensuring that senior water rights are not impinged upon by use of surface water to develop a CSP facility.

7.6.4. Local Representatives and Members of Economic Development Groups

Local groups have authority over positions that they will take regarding local CSP development and transmission lines in their areas. Sometimes their positions are formalized in permanent groups that are set up to advise federal agencies or other organizations in their decision processes, especially where public comment is mandatory. Such groups do exist in the San Luis Valley.[47]

7.6.5. Colorado State Land Board

One of the interviewees indicated that there has been some interest in developing state lands. A 2007 Colorado statute allows state lands to be used for solar energy facility siting. If CSP facility siting or transmission siting were proposed on state lands, the State Land Board (SLB) would be involved. The SLB, a branch of the Colorado Department of Natural Resources, manages its 3 million acres of land and 4 million acres of mineral rights for the benefit of eight trusts. The eight beneficiary trusts are: (1) School lands, (2) Land for public buildings, (3) Land for penitentiary, (4) Land for University (University of Colorado), (5) Salt springs — saline lands, (6) Internal improvements, (7) Land for agricultural university (Colorado State University), and (8) the Hesperus trust (the Fort Lewis grant). The board is charged with generating a reasonable and consistent income for its trust beneficiaries and making its decisions based on the long-term best interests of the beneficiaries.

The State Board of Land Commissioners, established in 1876, oversees the stewardship of the use of Colorado's land, from agriculture to mineral exploration to general significant revenue for its trust beneficiaries through agricultural leases for grazing and crop lands, mineral development, and interest earned on invested funds.

In recent years, the board has expanded efforts to increase revenue through commercial development activities and leasing land for recreational purposes.

The land board is entering leases for State Land Board land. The State Land Board is working on a model solar lease contract for land to be used for solar development.[48]

7.6.6. Colorado Department of Transportation

A significant amount of property is available for different types of utilities and proximity to highways.

The railroad is often the straightest line and railroad rights-of-way (ROW) can carry transmission lines and power poles.

7.6.7. Colorado State Treasurer's Office

This office was mentioned as a key decision-maker, but its role was not stated.

7.6.8. Transmission Planning Groups

A thicket of groups is working on renewable electricity transmission issues. As explained by one outside-the-Valley stakeholder:

> The Western Governors' Association has a process underway called the Western Renewable Energy Zone process, WREZ, and it's a process to identify transmission areas for renewable resources. And then there are numerous other transmission groups where transmission planning is actually carried out, such as Colorado Long-Range Transmission Planning Group, CLRTPG. There's a forest of acronyms here.
> There is the CREPC, the Committee for Regional Electric Power Coordination, I think, you can Google CREPC and get the expansion. There is the TEPPC, Transmission Expansion Power Planning Council, and there are many more acronyms. In a sense, I say this just to kind of illustrate that there are so many things going on which really makes true coordination difficult because you have each of these venues answering to different facilities and within each of these organizations also you have various utilities, you have investor-owned utilities, you have rural cooperatives, you have municipals, you have federal power agencies, you have tribes, you have a whole gamut of entities, all of whom answer to different regulatory and oversight bodies. Investor-owned utilities have to answer to their regulatory commission, the co-ops answer to the Tri-State's board and their members, municipals answer to city councils; tribes, of course, have their own governments, and so it makes transmission planning very difficult. It is a very complicated issue which has to be resolved soon, very soon, because we need to get wind energy from Wyoming to Los Angeles, we need to get solar power from Arizona to Los Angeles, we need to get...in a nutshell, we have a great need for aggressive interstate transmission of renewable power.

7.7. State Legislation

Two Colorado statutes, both enacted in 2007, have particular relevance for the progress of CSP development in the SLV: (1) S.B. 07-100 and (2) S.B. 07-091.

7.7.1. S.B. 07-100

This statute, concerning measures to ensure the adequacy of Colorado's electricity transmission requiring utilities to designate energy resource zones, went into law on March 27, 2007. It streamlined transmission approvals by basically allowing the investor-owned utilities to site new transmission before projects were developed. Previously, transmission lines could not be built without a project at the other end. An outside-the-Valley stakeholder said:

> ...Last year...we got legislation to double the RPS and to streamline the transmission approvals. That bill number was SB-100, which streamlined transmission approvals basically by allowing the utilities—the investor-owned utilities in this case—to site new transmission before projects were developed. Previously, the transmission couldn't be built without a project at the other end, but because transmission can take five years or more to build, and a new wind project can take one year, a new CSP project can take two or three years, compared to the five years or more that transmission takes, it was a chicken and egg thing. We just couldn't get new transmission sited, yet without transmission new projects couldn't be developed. It was a real chicken and egg problem which SB-100 began to help overcome.

7.7.2. SB 07-091

The bill, enacted into law May 29, 2007, created a task force to identify the state's renewable areas.[49] An outside stakeholder said that existing transmission in the Valley would be sufficient to handle the transmission needs of a 200-MW CSP plant. This stakeholder said:

> Yes, but that would pretty much exhaust, as I understand it...and also, again, I would refer you to work with GEO and others on Senate Bill 91 and Senate Bill 100 where they were looking at resource zones, especially the Senate Bill 91 report. This is the GEO which developed a map of the places where there are the best opportunities for renewable energy so that people can start planning transmission to reach those areas. The San Luis Valley obviously scores as the best central solar in the state and so the utility is planning for an upgrade of transmission east out of the valley. Tri-State and Public Service Company have talked about joining forces to build a line out.

Subsequent to the publication of the S.B. 91 report, the GEO issued another report on transmission planning,[50] called the Renewable Energy Development Infrastructure or REDI report. This is the report referred to in the quote. [51] The report shows the planned transmission line from the Walsenburg Substation to the proposed Calumet Substation, and then over La Veta Pass into the San Luis Valley. The line crosses the Trinchera Ranch, but almost entirely avoids Costilla County. Another stakeholder pointed out that there could be federal NEPA requirements if a plant were hooked up to a transmission line that crosses federal lands. One stakeholder commented that if the PUC did not accept the transmission line proposed by Xcel Energy and Tri-State as shown in the REDI report, then the two utilities would have to move the location of the lines and deal with the *Western Area Power Administration (WAPA)* who owns land in the vicinity.

An outside stakeholder said that SB 91 has identified 26 GW of undeveloped renewable energy in the state. He said:

> We should be looking at green electrons as an export commodity from this state, just as we do wheat or cattle, etc. To do that, we are going to need some commitments for infrastructure for transmission.

Another outside stakeholder described Sections 123's and 124's bearing on CSP development:

> The 124 has no bearing on concentrated solar power, but it is often confused because to a layman who might ask well why is Xcel Energy installing all of this PV and no CSP, the difference is that Xcel does not own nor do they make money in photovoltaic. The difference is that rather than utilizing the 2% retail rate impact tax that is authorized in section 124 (that is the no more than 2% of our monthly bill could be increased to fund the portfolio standard) CSP is not contained within that 2%, nor is the solar carve-out at all related to concentrating solar power, so the solar carve-out is for photovoltaic. That's 4%.
>
> Interviewer: 4% is what?

Stakeholder: 4% is approximately ... 90 ... MW by 2020. It is important to understand the structure of 123 versus 124 if you are really to understand how CSP might happen from a statutory and regulatory standpoint in the San Luis Valley....
...One of the things... [to] look into is the property taxation structures for large solar plants. Currently they are taxed upon the market value of the system. They are taxed upon the value per rated megawatt of the system. What Senator Schwartz has looked into is what if that property taxation rate was changed to be a commodity basis, similar to the way agricultural land is taxed; for example, the productivity of the land, you get so many bushels, for example. Essentially what she has investigated is taxation, and in fact there has been a bill on this, Senate Bill 177, that modifies the taxation structure for large-scale solar systems. It actually starts the property taxation lower than they currently are in the earlier years, but the amount of property tax generated would go up over time as their value of that commodity goes up. So as the megawatt hour becomes more valuable at the time, that megawatt hour is taxable from a property tax basis.

Interviewer: So that would affect the county.

Stakeholder: Right. Exactly. More importantly, it brings a greater guarantee of longterm revenue to the county whereas the current taxation structure is continuously decreasing based on the value of the system. These systems are taking advantage of accelerated depreciation so they are depreciating very quickly. I think that is a very important piece that Senator Schwartz has hit on. It is unclear what the impact would be to a developer.

PART TWO: THE SLV TRANSMISSION CONTROVERSY

8. THE SLV TRANSMISSION CONTROVERSY

Summary

Within a federal and state policy context broadly supportive of utility-scale solar development, in May 2009, Xcel Energy and Tri-State Generation and Transmission Company jointly filed for a Certificate of Public Convenience and Necessity (CPCN) to build a 95-mile high-voltage transmission line between Walsenburg and Fort Garland, Colorado. This action sparked a public policy debate over preserving the Valley's scenic and wildlife resources versus the need to transport renewable electricity to the Front Range and to increase power reliability in the Valley.

The dispute over the proposed placement of the SLV-Calumet-Comanche line, planned to be operational by May 2013, illustrates some of the difficulties and challenges facing large-scale renewable power development in rural and scenic areas in the United States.

The CPCN filing was met with immediate legal opposition from Valley stakeholders—notably the Trinchera Ranch, a privately held wildlife conservation and hunting preserve.

Other Valley stakeholders testified for and against the transmission line at public hearings held in the area. In Denver, the PUC conducted formal hearings on the CPCN docket, and a decision is expected toward the end of April 2010. The losing side will likely appeal the decision.

8.1. Introduction

This section of the report briefly addresses the transmission controversy that erupted in the spring of 2009 after the stakeholder interviews reported in Part One were completed.

The debate was fueled by the Xcel Energy and Tri-State Generation & Transmission Company (Tri-State) filing in May 2009 for a CPCN to build a 230-kV transmission line between Walsenburg and Fort Garland over LaVeta Pass, a distance of 95 miles.

The rationale for building the line was to carry out federal and state policies fostering the development of large-scale renewable energy technologies by bringing transmission capabilities to resource-rich areas like the SLV. Significant policy development at the federal and state levels had laid the groundwork for a transmission line proposal. Federal policy developments described in this report's Introduction and Background are mentioned and relevant developments in Colorado policy are summarized. The positions of Xcel Energy and Tri-State and of the opponents to the transmission line are summarized.

8.2. Policy Context

Federal and state policies favor utility-scale solar development.

8.2.1. Federal Policy

As noted earlier, Congress has passed legislation and the federal government has adopted strong renewable-energy development policies. Among these are the following federal policy actions.

8.2.1.1. BLM Policy Memorandum

On April 4, 2007, the then Acting Director of BLM James M. Hughes issued an instruction memorandum on Solar Energy Development Policy, citing the National Energy Policy Act of 2001 and the Energy Policy Act of 2005 (P.L. 109-58, August 8, 2005).

These statutes encouraged the development of renewable energy resources including solar energy.

Specifically, Section 211 of the 2005 EPAct encourages the approval of at least 10,000 MW of non-hydropower renewable energy projects on the public lands within the next 10 years.

8.2.1.2. BLM-DOE PEIS

As noted earlier, BLM and DOE have joined forces to develop a PEIS to aid in streamlining permitting processes for large-scale solar projects on the public lands. The SLV has huge areas of public lands (nearly 3 million acres), and four areas of the Valley have been federally designated as official solar energy study areas.

8.2.1.3. Investment Tax Credit (ITC)

As noted earlier, in 2008 the federal ITC for solar development was expanded to include utilities and extended to 2016. This provides a key financial incentive for large-scale solar development by helping to reduce their cost.

8.2.1.4. DOE Report to Congress

As described earlier, the DOE reported to Congress in 2007 that the solar resource and available suitable land could support generation of up to 6,800 GW of electricity. Based on geographic information system (GIS) and solar resource analysis conducted by NREL, the SLV has been identified as a prime national solar resource.

8.2.2. Colorado Policy

In 2004, Colorado voters passed Amendment 37 to the Colorado Revised Statutes, calling for a renewable energy portfolio standard (RPS) of 3% beginning in 2007, and incrementing to 6% in 2011, and then 10% in 2015 and beyond.[52]

The renewable requirement included a "solar set-aside" stating that at least 4% of the renewable generation in each year must come from solar resources. Xcel Energy met the first RPS 8 years ahead of schedule.

In 2006, then-candidate for Governor Bill Ritter campaigned on a "new energy economy" as one of his three key policy themes. Once elected, the governor established an energy office—the Governor's Energy Office (GEO)—to carry out his energy agenda. In 2007, Governor Ritter signed into law House Bill 1281 that doubled the RPS requirement (to 20% in 2020) and on March 22, 2010, the governor signed into law H.B. 10-1001 increasing the RPS requirement to 30% by 2020.

In December 2009, the GEO released the REDI Report calling for actions to meet a benchmark goal for meeting Colorado's RPS and reducing carbon dioxide emissions in Colorado by 20% by 2020.53 In summary, the report's major recommendations for action included the following.[54]

- Greatly increase investment in demand-side resource (energy efficiency, demand-side management, demand response, and conservation)
- Greatly increase investment in renewable energy development, particularly utility-scale wind and solar generation.
- Accelerate construction of high-voltage electric power transmission to deliver renewable energy from Colorado's renewable resource generation development areas to the state's major load centers.
- Strategically use natural gas-fired power generation to provide needed new power to the grid and to integrate naturally variable renewable resources.
- Consider decreasing the utilization factor of coal-fired generation and/or consider early retirement of the oldest and least efficient of the state's coal-fired generation stations.

The report noted that "delays associated with siting and permitting of transmission lines will hamper Colorado's utility-scale renewable energy development unless modifications are made to the process."[55] Indeed, the report appears to advocate a state-level transmission permitting framework to make local-area transmission siting decisions, citing a 2009 Memorandum of Understanding (MOU) between the WGA, the DOI (Interior), the USDA (Agriculture), and DOE (Energy).

This MOU creates assistance to the WGA's Wildlife Council, working with member states, to create state-based decision support systems for such siting decisions. The effort will support development of "effective land uses" while providing "healthy and productive landscapes."[56] The purpose of the MOU is to "help the renewable energy industry reach its goals" while also "preserving prairie and mountain ecosystems" in Colorado.

8.3. The Current Situation

Although the ambitious goals of the Colorado RPS could be met partially by distributed (customer-owned) generation, it is widely (but not unanimously) agreed that larger-scale solar and wind projects in Colorado's rural areas will need to be built as well (to meet the higher RPS).

The REDI Report focuses on how Colorado can "most effectively address the challenge of building new high-voltage transmission lines to deliver utility-scale renewable power from Colorado's rich renewable resource generation areas to the market".[57] What the Valley lacks is adequate transmission infrastructure to bring that resource to market. Local opponents in the SLV argue that building huge solar arrays and massive power lines to export renewable energy would diminish other environmental values innate to the area. They argue, moreover, that approving and constructing large transmission lines before the generation capacity has been put under contract is imprudent, calling it the "field of dreams" approach. The GEO and the utilities from building projects in Colorado and that "if you build it, they will come."

Tri-State's customers in the Valley are served by only one transmission line (Tri-State has roughly a 100-MW load in the Valley; Xcel Energy has a 25-MW load in the Valley). For Tri-State, the primary purpose is to solve a critical need to improve the electric service to the Valley and more reliably serve their SLV customers. Currently, if a single outage event were to occur on the existing Poncha–San Luis Valley (230-kV) transmission line, the remaining system would not be able to supply enough power to meet the peak loads in the area. These include agricultural water-pumping loads, which are critical for growing crops.

Xcel Energy and Tri-State jointly filed for a Certificate of Public Convenience and Necessity (CPCN), seeking permission to build the proposed power lines (and new Calumet substation near Walsenburg). This application opened a Docket before the Colorado PUC in which the merits and shortcomings of the proposed transmission assets, and the routes of the transmission lines, were disputed before the Commission.[58] Table B-1 in Appendix B highlights the major regulatory events and project milestones. Thus far, there have been four rounds of testimony filed before the Colorado PUC. A final decision, likely to be appealed, was expected by the end of April 2010.

The proposed line would require a 150-foot wide corridor and latticed towers up to 120-feet high. It would cross La Veta Pass traversing pristine valleys and mountains as it passes by Great Sand Dunes National Park, crossing as well areas designated as prime areas for wind and solar power development. The exact route has not been selected, and does not have to be selected for the CPCN to be granted. If the utilities get the CPCN, they can, if they must, legally condemn private lands to build the lines.

In their application for the CPCN, Xcel Energy and Tri-State requested a determination of permissible noise levels for electric transmission facilities "notwithstanding the maximum permissible noise levels established [elsewhere in law]".[59] The utilities want this threshold defined by the Commission in order to avoid civil action claims in the future.

A partnership by Xcel Energy and Tri-State Generation to build the (estimated) $180 million project to export solar energy (to Colorado load centers) and increase reliable electricity (in the Valley and in south-central Colorado) is supported by a number of stakeholders around the state, particularly proponents of renewable energy, the Governor's Energy Office, and the renewable energy industry, but faces opposition from a number of opponents as well, including a wealthy, local landowner and some environmental organizations based in the Valley.

The landowner, fearful of the visual impact the line would cause as it crosses his 171,000-acre Trinchera Ranch, has proposed an alternate plan that would upgrade existing power lines and add a new proposed high-power line over Poncha Pass. A federally designated corridor already exists at Poncha Pass for future energy lines, connecting to Cañon City. All of the existing transmission lines—one 115-kV line and two 230 kV-lines—come over Poncha Pass. According to the utilities, for truly reliable service, all transmission resources connecting the Valley to the Front Range should not be routed through the same corridor. The utilities argue that the proposed alternative is not acceptable because it fails to create the type of redundancy that utilities strive for in most transmission applications. One large forest fire or winter storm could take out the Poncha Pass lines and leave the valley without electricity.

8.4. Xcel Energy and Tri-State Perspectives

Xcel Energy and Tri-State G&T intend to co-own the proposed transmission assets, although they expressed different reasons for needing them.[60] Xcel Energy (also known as Public Service Company of Colorado) proposed the line to stimulate the development of large-scale solar energy (both concentrated PV and solar-thermal plants) in the SLV, also referred to as Energy Resource Zone (ERZ) 4. In his testimony filed before the PUC, Gerald Stellern, manager of transmission assessment management at Xcel Energy, wrote that "the Project meets the SB-100 criteria to alleviate transmission constraints to ERZs 4 and 5, and it will allow generation resource additions to meet Public Service's (renewable energy) needs... and can accommodate approximately 1,500 MW of new generation."[61]

Xcel Energy pushed an amendment to the bill in June 2009 that would give the utilities the right of first refusal (RFR) before CEDA could proceed with financing projects. The CEDA Board is unanimously opposed to this amendment. If the amendment passed, the effect could signal the demise of CEDA and future opportunities for its financial support for projects that transmit clean energy. The CEDA Board is opposed to the amendment because:

- Independent transmission developers seeking CEDA financing would not participate because they would not want to reveal proprietary development information to utilities.
- A utility RFR could subject CEDA to unnecessary delays.
- A utility RFR could have the effect of an exercise of undue market power control, potentially leading to an anti-competitive outcome.
- A utility RFR could lead to an exercise of undue control by a utility over a public body composed of individuals appointed by state elected officials.

S.B. 100 drove the process of identifying ERZs around the state and potential transmission lines that would need to be built. Regarding public outreach associated with S.B. 100, Stellern stated that "Public Service has held numerous meetings, workshops, and technical sessions to plan the S.B. 100 projects by communicating the information to all stakeholders. The result of that work was five Energy Resource Zones and eight projects that were presented to the Commission in November 2008 in an information filing."[62] In its rebuttal testimony, Xcel Energy responded to some of the arguments presented by other stakeholders before the PUC.[63] Karen Hyde explained that Xcel Energy cannot agree with Western Resource Advocates' (WRA) position that the proposed project can "provide access only to renewable resources." She also discussed why the positions held by the interveners for Trinchera Ranch, that the transmission lines "should only be built to accommodate generation under contract [in the Valley]... are short-sighted, imprudent, and unlikely to meet Public Service's solar resource plans over the next several years." Xcel Energy argued that alternative plans and routes proposed by Trinchera Ranch do not look out far enough into the future. "The potential solar resources in the San Luis Valley are in the thousands of megawatts, not hundreds. Artificially limiting transmission to the bare minimum necessary to connect resources that result (only) from the 2007 Resource Planning Docket would not be wise."[64] She explained why the transmission lines have to be built ahead of the generation market—that is, before firm commitments to develop solar energy in the Valley are under contract. "The availability of transmission capacity is watched carefully by bidders; they will look to develop projects in areas where it is likely that they will have an economic advantage over bidders in other areas." Xcel Energy also disagreed with a study done by engineers hired by Trinchera Ranch that describe alternative routes through Poncha Pass, thus avoiding the Ranch. "We believe that the proposed southern transmission route makes the best sense from reliability and quite frankly a common-sense standpoint," said Xcel Energy spokesman Mark Stutz.[65] A new Poncha Pass line would require an extra 60 to 70 miles of line at about $1 million a mile. Tri-State, which provides electricity to rural electric cooperatives, including the SLV REC, has asked the federal government to secure financing for the transmission line. This request triggered prescriptive public proceedings in the local area. The Rural Utilities Service (RUS) decided that Tri-State must complete a more complex environmental impact statement rather than a simpler environment assessment on the impact of the proposed project transmission line on the environment."The project sponsors announced that the Rural Utilities Service will require an environmental impact statement, a more exhaustive level of review than the utilities planned for. The Rural Utilities Service is the federal agency that funds rural utility projects and is considering loaning $82.3 million for this project."[66]

8.5. Trinchera Ranch Perspective

The Trinchera Ranch in the San Luis Valley changed hands in 2007 from Steve Forbes to billionaire hedge-fund manager Louis Moore Bacon. Forbes said he had sold the 171,000-acre ranch near Fort Garland to Bacon because Bacon had a solid conservation record and could be trusted to preserve the ranch.

The Pueblo Chieftain said Bacon paid $175 million for the ranch, which has views of 14,000-foot-high peaks one of the highest prices ever paid for a ranch. The Forbes family had held the ranch, 160 miles south of Denver, for four decades.[67]

Bacon had set up the Moore Charitable Foundation in 1992 to aid nonprofit groups that focus primarily on conservation and the protection of natural resources.[68] "Louis Bacon has passionately devoted much of his life and resources to the protection of extraordinary properties," Forbes said. "By finding such a committed owner, we are certain Trinchera will thrive and be enjoyed, as it is, for years to come." A spokesman for Bacon called the ranch an extraordinary property for its scenic grandeur and unspoiled natural habitat.

Bacon strongly opposes the proposed power line and has hired lawyers, engineers, and an independent transmission analyst to fight it. The Ranch filed testimony and developed a report proposing alternatives to the currently proposed La Veta Pass route, which crosses the ranch. Among the proposed alternatives are:

- Upgrade existing lines and a transformer
- Add 150-MW of solar energy with power storage to reliably service the Valley
- Add a 230-kV line north over Poncha Pass.

"All these alternatives meet the reliability need, allow for getting solar out of the Valley and are less expensive than the proposed line," said Cody Wertz, a spokesman for Trinchera Ranch. "Xcel and Tri-State have not shown there is a demonstrable (reliability) risk over Poncha Pass," Wertz said.[69]

In his testimony before the PUC on behalf of Trinchera Ranch and Blanca Holdings, Michael McFadden argued that "lower cost alternatives will accomplish the same goals as the facilities proposed by the Company."[70]

Moreover, he argued that "there are no commitments or assurances that generation capacity sufficient to justify the proposed transmission facilities will be constructed and connected to the system.

Without these assurances, these facilities cannot be considered used and useful." He concludes his argument by suggesting that the PUC should "place the risk of failing to attract sufficient generation on the shareholders, not the customers."[71]

8.6. Local Environmental Perspectives

The Colorado PUC held public hearings in the area (on Nov. 9, 2009, in Walsenburg, and Nov. 10, 2009, in Alamosa) to receive public commentary from local stakeholders. Two SLV groups—the San Luis Valley Ecosystem Council and Colorado Open Lands—offered pertinent comments opposing the proposed lines.

The San Luis Valley Ecosystem Council strongly criticized the proposed line. Its comments emphasized the benefits of distributed generation and the environmental impacts transmission lines and new bulk-power generation facilities would likely have on local ecosystems:

> Study after study concludes that distributed, or small-scale, rather than concentrated or industrial-scale, generation projects are more cost effective, more efficient and much quicker to bring online. Studies also show that small-scale projects provide long-term financial benefits to local communities through sustainable jobs, robust returns on investment for local businesses and homeowners, and reduced carbon footprints.

> Furthermore, energy that is produced in the place where it is used is much more efficient since it eliminates transmission losses and large-scale production inefficiencies. Without question, distributed generation projects have dramatically less impact on the environment.[72]

The council believes that system reliability and voltage loads would best be served by creating a locally generated source of electrical power supply and transmission.

> We already have the existing infrastructure through our 30 odd sub-stations and center pivot corners (for example). It could come online more quickly than a transmission line and be much more cost-effective for consumers. This self-sustaining alternative deserves to be given thorough analysis.

Several important historic resources are located within the proposed transmission corridor plans. These include the Sangre de Cristo National Heritage Area and the Los Caminos Antiguos Scenic By-way. Because of these two special designations, the council believed that it is important to preserve the visual landscape and that the transmission line would diminish the historic integrity of the area.[73] Moreover, Mt. Blanca is considered one of four mountain peaks in the Four Corners area to be sacred among various tribes who inhabited and traded in this area. The council argues that tribes should be consulted as an integral part of this public process.

According to the council, habitat fragmentation; threatened, endangered and sensitive species habitat; as well as critical cores and linkages for wildlife habitat will be impacted by the proposed transmission line. Wildlife habitat fragmentation caused by transmission lines (including branch power-lines), pipelines (including feeder pipelines) and roads occur both during and after construction. As such, wildlife habitat must be examined on an individual project and site-specific basis. The only way to accomplish this requirement is to ensure that the transmission corridor is spatially evaluated for direct, indirect, and cumulative impacts.

Proponents and opponents alike spoke at PUC public hearing on November 9th (Walsenberg) and 10th (Alamosa). Considerable concern was expressed that large central-station solar power facilities would consume precious Valley water. Residents suggested the power lines be buried, but the utilities said that would increase the cost ten-fold. Adding to residents' consternation, they learned that the utilities could use eminent domain to seize rights-of-way from property owners unwilling to sell.

The utilities issued a "route refinement," eliminating the route through the rural La Veta Acres subdivision 10 miles west of La Veta, instead proposing to run the lines far to the north. The announcement of the route change "shocked and delighted opponents," including Public Outcry Over Stealing Scenic Excellence, a still-forming group of about 60 residents who had planned to urge the utilities to adopt a different route.[74]

In criticizing the transmission plan, residents claimed they were not notified about the August 2009 public hearing. They claimed that nearly 40 homes and historic buildings were omitted from the initial survey for the line. Tri-State acknowledged that there were problems with maps the utilities used to identify the originally proposed routes. "There were instances where there could have been properties and structures that were not appearing on the maps we used," Jones said. The new proposed route would not add to the cost of the project, nor would it impact any residential areas as dense as La Veta Acres, he said.[75]

8.7. Western Resource Advocates Perspective

Western Resource Advocates, an environmental law firm with a history of intervening before the Colorado PUC, filed testimony in support of the line, but with a few conditions to ensure that the lines only serve renewable energy projects and that environmental impacts are limited.

In WRA's Surrebuttal,[76] Tom Darin commented that "the importance of WRA's proposed conditions on each CPCN," and responded to the applicants' comments and objections with regard to the proposed conditions for each CPCN. The issues fell into three primary categories: (1) demand-side management (DSM) and distributed generation (DG); (2) the use of the proposed transmission lines to facilitate the development and delivery of renewable energy, as described in the applications; and (3) the responsible mitigation of visual impacts, and the importance of early consideration and planning that is necessary to be sure that visual impacts are properly incorporated into transmission siting and design."

Mr. Darin wrote that the utilities' objections to WRA's proposed conditions generally fell into these five categories:

1. DSM, DG, and environmental concerns, such as visual impacts, are not relevant in a CPCN proceeding.
2. Reporting on DSM and DG is either redundant or over-burdensome, and the commission's interest in such information would be beyond its jurisdiction as it relates to Tri-State.
3. Applying a rebuttable presumption to future generation facility CPCN applications before the commission would frustrate FERC's open access requirements.
4. Requiring a percentage of the proposed transmission line's capacity to be committed to renewable energy prior to commencing construction of the proposed facilities might pose implementation challenges based on Xcel Energy's experience relative to a similar condition in a CPCN hearing in Minnesota.
5. WRA seemed to single out visual impact as a key area for mitigation.

8.8. Perspectives of the Colorado Division of Wildlife and Colorado Open Lands

The conservation group Colorado Open Lands and the Colorado Division of Wildlife (CDOW) voiced concerns about the impact a transmission line could have on scenic and wildlife resources.

The 140-mile line would have "a negative impact on the land and resources," Daniel Pike, president of Colorado Open Lands, said in his testimony.[77]

The CDOW also commented on the potential environmental impacts of each segment of the proposed lines. Identifying the best corridors is "probably the most important factor for CDOW."[78]

Southwest Regional Manager Thomas Spezze wrote that from Walsenburg to La Veta Pass, "our preference is for corridor segments C, E, G, and H because they have the fewest additional impacts.

This route uses an existing right-of-way where large transmission lines already exist" (to minimize the impacts on elk, deer, and antelope winter ranges).[79]

Avian concerns were a much more pressing issue to the CDOW. "There are numerous migratory birds that use what is considered a triangle of very important wetlands, grain fields, and roosting areas to the north, south, and east of the proposed corridors." Also, during the spring and fall of every year, thousands of Sandhill Cranes will use these areas for feeding and roosting.

Pelicans will also migrate between the San Luis Lakes, the Alamosa Wildlife Refuge, Smith Reservoir."

To avoid avian transmission line collisions, "we recommend marking all lines, especially the top wire, throughout the entire project.

CDOW recommends that Tri-State install Yellow Swan Flight Diverters on overhead static wires. It is possible to stagger Swan Flight Diverters to minimize the number of devices required and cost."

CDOW's primary concern regarding aquatic species "is to protect riparian habitat along all streams, reduce erosion and sedimentation by minimizing stream crossings, and protect Rio Grande cutthroat waters."[80]

In all corridors, CDOW "recommends riparian habitat protection to minimize sedimentation and erosion by providing a minimum of a 50-foot no disturbance buffer zone on each side of the stream, and avoiding surface disturbance within 300 feet of the riparian zone to the maximum extent practicable."[81]

8.9. County Government Perspectives

The proposed transmission corridors go through Alamosa, Costilla, and Huerfano Counties. Some county commissioners were favorable. In testimony at a local PUC hearing, commissioners for each affected county supported building the transmission line."We are in favor of the transmission line and more solar energy in the valley," San Luis County Commissioner Darius Allen said.[82] The line was expected to ensure electricity supplies to power the 3,000 sprinklers used by Valley farms. "If we have the power interrupted, it would be devastating," Allen said. Rio Grande County Commissioner Doug Davie said that losing power on a 120-acre sprinkler could curb a potato yield by seven sacks an acre per day, worth about $5,400.[83] County commissioners clearly anticipate financial benefits. Unless WGA recommendations are adopted, the utilities will be paying property taxes to the counties for relevant portions of the lines. In addition, jobs could be created if personnel are hired to maintain the line.

Part Three: Discussion, Conclusions, and Recommendations

9. Discussion, Conclusions, and Recommendations

US public policy fosters increasingly widespread use of renewable energy. The purpose of a stakeholder analysis is to gain understanding of the ways in which stakeholders perceive renewable energy technology (in this case, CSP) and how it can best be used. Do they understand it? What is the expertise that local stakeholders can bring to Colorado energy decision-making processes? The point of the analysis is to provide information that helps to improve the process of renewable energy siting in Colorado.

The research stance taken in this study is one of neutrality in what has become a conflicted community. The range of opinions about the positives and negatives of CSP development in the SLV was identified qualitatively. The purpose is to describe the perceptions among stakeholders within and outside the Valley, and to hold up a mirror on the viewpoints contained within the decision process. Most of the key issues are likely represented in these report's findings.

Better ways to site renewable energy transmission lines are needed. Other renewable resource transmission controversies are occurring elsewhere in the country, one in southern California. "Facing enormous costs and fierce opposition from environmental groups, the Los Angeles Department of Water and Power (DWP) announced on March 10, 2010, that it had dropped plans to build an 85-mile-long 'green' power transmission line across desert wilderness preserves and scenic ridgelines. The announcement reflected a shift in policy toward developing renewable resources closer to the DWP's existing power corridors. The DWP submitted a right-of-way grant application to the US Forest Service in 2007 for the project, designed to bring electricity generated by solar, geothermal, wind and nuclear power to Los Angeles from the southeastern California deserts and Arizona. Environmental and community groups were outraged by the DWP's plans to route high-voltage lines and 16-story towers through the Big Morongo Canyon Preserve north of Palm Springs, Pioneertown near Yucca Valley, Pipes Canyon Wilderness Preserve and a corner of the San Bernardino National Forest before linking with existing DWP power lines in Hesperia."[84]

This part of the report provides summary review of how the Valley situation developed, answers the research questions raised by the Research Advisory Committee identified in the report's Introduction and Background, discusses policy actions proposed by solar advocates in light of the study's findings, presents broad conclusions, and recommends state policy and research actions.

9.1. Summary Review

The early signs of opposition to the transmission were already evident in October 2008 and probably sooner. The opponents made no secret of their concerns, stating in public meetings that although they favored distributed solar development, they opposed a second transmission line into the Valley over La Veta Pass. They raised a number of concerns publicly, including the industrialization of the Valley, the large amount of land required, and the water needed for a large-scale solar facility. Other stakeholders within the Valley were aware of these opponents and their concerns. Stakeholders from outside the Valley who had visited and attended meetings there were also aware of these individuals and groups and the nature of their concerns.

Nevertheless, the responsibility for attending to local concerns apparently rested with the private independent power producers (IPPs) who were submitting bids to Xcel Energy to produce renewable electricity. As part of their bids, these developers were expected to vouch for the fact that they could deliver the electricity to the utility's transmission lines. This meant that they had obtained, or expected to obtain, local permits to site their facilities.

The Xcel Energy RFP for bids for solar power went out as a direct consequence of Colorado's renewable portfolio standard (RPS) that resulted from the passage of citizen-initiated Amendment 37 and subsequent state legislation. The latest RPS legislation (H.B. 10-1001, March 2010) mandates that 30% of retail electricity produced by investor-owned utilities must come from renewable resources. A minimum of 1.5% must be from onsite generation and 28.5% from utility-scale projects.

Although those in control of the bureaucratic process and the IPPs may have understood the relevant legal actions they had to take for their desired outcomes, many local stakeholders— whose neighborhoods would be impacted—were confused about those decision processes and were sometimes frightened by unknown technology, economic risks, and outside organizations with the clout to seriously affect them. Typically, they did not have the expertise (nor necessarily the wherewithal to purchase it) that could provide them with third-party technical advice on the potential economic, environmental, water, land, societal, and other impacts of such developments, but had to rely on the testimony of those organizations responsible for, and seemingly wedded to, CSP development.

The process took on an adversarial tone when Xcel Energy and Tri-State filed for a Certificate of Public Convenience and Necessity (CPCN) with the Colorado Public Utilities Commission to build the San Luis Valley-Calumet-Comanche transmission line over La Veta Pass. As was summarized in Part Two, lawyers represented both sides of this issue.

Although organizations provided public briefings on their plans and posted information and maps on the Web, no one appeared to have legal responsibility for educating the citizenry or mitigating potential adverse environmental or societal impacts.

9.2. Research Questions and Answers

As described in the Introduction to this report, the study's Research Advisory Committee asked several key research questions. These questions are answered below with brief summaries of findings from the study.

- *What is the level of awareness and knowledge of key SLV stakeholders?*

The need for education was a familiar theme in the interviews. Many of the stakeholders believed that the local community should be educated so that they understand more about why CSP development is necessary; how it contributes to the goals of the federal and state policy communities; how it helps the environment; how the Valley needs more electricity; why its electricity supply is vulnerable; why distributed solar PV cannot provide all of that electricity and security; what the impacts would be on their communities, landscape, and water; and what benefits—especially economic benefits—the Valley populace would enjoy as a result of the development.

- *What is the level of community support and opposition?*

There is a high level of support in the Valley for renewable energy development. Many Valley stakeholders appear to favor solar development. The Valley's lengthy history of grassroots activism and support for passive solar buildings, solar thermal, and PV seems to inform this solar consciousness. Added to that is the successful integration of the 8.2-MW Sun Edison PV plant sited near the town of Mosca. However, some stakeholders believed that most SLV residents do not understand CSP. Those characterized as supportive included the county commissioners, farmers with land to sell, those working on economic development in the Valley, the college, and those concerned about workforce opportunities. Those opposed were (accurately or not) identified as water interests, environmental groups, the Trinchera Ranch, residents of Crestone (a Valley community near the Sand Dunes National Park and Preserve), the rural electric cooperative, and Spanish land-grant families in Alamosa and San Luis.

The perceived benefits of CSP development can be classified into five categories as follows.

1) *Economic benefits* (potential improvements in the SLV economy, job creation, provision of tax benefits to local counties, trapping dollars in Colorado's economy)
2) *Environmental benefits* (puts lands and solar resources to best use, reduces greenhouse gas emissions; addresses climate change, helps "save the planet," is "clean energy")
3) *Technological benefits* (improves security and diversity of power supplies, reduces dependence on imported energy, provides dispatchable energy)
4) *Social-psychological benefits* (solar development can become a source of pride, positively affects the community identity)
5) *Policy benefits* (the community wants RPSs to be met).

Some differences between the perspectives of stakeholders within the Valley and those outside the Valley were observed. The responses of inside-the-Valley stakeholders gave greater emphasis to environmental and social-psychological benefits, whereas outside-the-Valley responses emphasized technological benefits (especially energy storage). Both groups emphasized the economic advantages, but inside-the-Valley responses focused more on the concern about the potential for a boom-bust cycle than did outsiders.

- *What are the key water, land, and other issues?*

Following is a summary of the concerns expressed in the interviews.

- *Environmental concerns* (the need for water for CSP development; land-use intensity)
- *Economic concerns* (concern that the Valley benefit economically, concern that a boom-bust cycle might occur from the CSP plant construction and that permanent jobs would be minimal)
- *Technological concerns* (concern that transmission would be needed to export CSP power; the technical efficacy of CSP was still in question; the intermittency of solar power was also a concern)

- *Social-psychological concerns* (concern that CSP is difficult to conceptualize; that it is difficult for the Valley people to deal with Xcel Energy)
- *Policy and regulatory concerns* (need for environmental impact statements; concern that regulation has not caught up with CSP development; lack of effectiveness of the 30% investment tax credit).

Local opposition to the San Luis Valley-Calumet-Comanche transmission line has organized and is waging a campaign to get the line stopped. The organized opposition, led by a local environmental group, is in contact with opposition groups in other parts of the country to share information and strategies on bringing proposed transmission projects to a halt.

- *Should there be local ownership of a CSP facility or should it be owned by CSP developers? What is the appropriate role of the local players?*

The answer depends on the perspective of the stakeholders. The SLV Resource Development Group believes that the local San Luis Valley Power Authority should own a CSP facility. The CSP developers believe that a CSP facility is too large, complex, and expensive to be locally owned. Local actors play key roles in ensuring local benefits and identifying negative impacts that must be mitigated. Generally speaking, the stakeholders outside the Valley want private ownership. Although some stakeholders support no export of power, others favor the revenues such exports would bring. The researcher's view is that the local players should have actual decision authority over the siting of CSP facilities on private lands. The part that distributed generation can play in providing electricity self-sufficiency to the Valley is currently unknown.

- *What are the key governmental entities involved in CSP decision-making and how can they work together?*

At the federal level, the key entities are the BLM, DOE, USFWS, RUS, FERC, and NERC. The USFS could also be involved. At the state level, the key entities are the state legislature, the PUC, and the GEO. Other state entities are also involved. At the local level, the county commissioners are the key decision makers, but the SLV Power Authority could possibly also be involved. The ways in which many decision-making entities could productively work together still need to be worked out; defining them is far beyond the scope of this study. Some preliminary actions are suggested in the policy recommendations in this report.

- *What are new models for the siting decision processes rather than the producer-buyerpurchase power agreement model? Can rural electric cooperatives play a larger role and, if so, how? Is there a model for developing power locally and using it where it is produced?*

Some Valley stakeholders have put forward a concept of meeting local power needs with a distributed system in which small solar installations at Valley substations would be complemented by distributed generation (rooftop PV and solar thermal) along with smart-grid technology so that the Valley could become energy self-sufficient. Their vision is of a locally owned solar network run by an SLV Power Authority into which local funds would be invested. The SLV Power Authority proponents believe that the local area has the financial wherewithal to develop and own a 100-MW CSP facility. The revenues from the electricity sales from this authority would be public revenues that could be used for schools, job training, welfare, and other services needed in this economically challenged region. This distributed-system idea is unlikely to work, however, unless the existing institutional structure for power generation and delivery is changed. The barrier is that the SLV REC must still purchase power from Tri-State under its long-term contract. The SLV Power Authority would not be able to sell its power to the REC for local distribution until someone bought out the Tri-State contract. The key question is: who would buy out the local utility's contract with power providers? The answer would probably have to be that the federal government would have to release the REC from long-term liabilities if it wanted communities to establish local area generation based on local sustainable energy. Otherwise, new sustainable energy would have to be exported to load centers to meet new demand.

- *What are the issues relative to transmission, such as crossing federal lands, environmental impact statements required, and public participation?*

The need for transmission was recognized by several stakeholders whose understanding is that the SLV needs to improve its aging grid infrastructure, which has been neglected for 30 years.

It was also believed that more electricity is needed in the Valley, but there was an emphasis that local demand should be met locally. Some stakeholders are interested in the ability to export electricity, thus bringing jobs and income into the Valley.

Concerns about transmission focused on concern about the "industrialization of the Valley floor," aesthetic concerns about the scenic attributes of the SLV (the 14,000-foot peaks of the Sangre de Cristo mountains and Mt. Blanca, the Blanca Ranch, and Trinchera Ranch areas; feared loss of control of decisions affecting the Valley; wildlife issues; and adversarial proceedings regarding the route of power lines.

The lines of local arguments can be succinctly summarized as follows:

- Should the SLV export power? Yes or no?
- (If yes), Should the line go over La Veta Pass or Poncha Pass?
- (If yes or no), To what degree should the SLV develop distributed energy systems?

There are proponents and opponents for each of these positions.

- *What are the policy, regulatory, and sociological barriers to CSP development?*

Although those in control of the bureaucratic process and the independent power producers understood the legal actions they had to take for the outcome they wanted to achieve, the local stakeholders—those whose neighborhoods would be impacted—exhibited confusion about those decision processes and were often frightened by unknown technology, economic risks, and outside organizations with the clout to seriously affect them. Typically, they did not have the expertise (nor necessarily the wherewithal to purchase it) that could provide them with third-party technical advice on the potential economic, environmental, water, land, societal, and other impacts of such developments, but had to rely on the testimony of those organizations seemingly wedded to, and responsible for, the CSP development.

- *What are the public lands involved in the Valley and what are the implications of CSP development on public lands? Is CSP included in the BLM's resource management plans? Are there concerns about sensitive wilderness areas?*

As part of its ongoing efforts to increase domestic energy production and ensure greater energy security, the Bureau of Land Management (BLM) initiated a joint Programmatic Environmental Impact Statement (PEIS) with the US Department of Energy (DOE) to assess the environmental, social, and economic impacts associated with utility-scale solar energy development on BLM-managed public land in six western States: Arizona, California, Colorado, Nevada, New Mexico, and Utah. The Notice of Intent (NOI) was published in the Federal Register on May 29, 2008.[85]

The BLM took this action after it became apparent that it had a backlog of applications for permits to develop large-scale solar energy projects on public lands but it could not issue permits because solar development was not part of its resource management plans. As the bureau works to develop a programmatic EIS that will streamline local areas, its personnel are climbing a learning curve about the potential land-use and other impacts of these projects. A final PEIS and Record of Decision is anticipated before the end of 2010.

- *What role are environmentalists playing?*

Local environmentalists are opposed to the proposed SLV-Calumet-Comanche transmission line. They favor distributed solar generation, including rooftop PV, in the SLV. Western Resource Advocates, an environmental organization based in Boulder, has intervened in the application for the CPCN to support the transmission line with certain limitations.

- *What are the institutional barriers to CSP siting?*

As noted, the decision processes are extremely complex, displaying a confusing mix of entities and procedures. Three siting decision models were identified (1) decisions by the SLV Power Authority, (2) county permitting, and (3) permitting on public lands. These processes are not well understood by the public, and not even by many of the Valley's stakeholders. As mentioned earlier, a Valley economic development group has fostered the creation of an SLV Power Authority that would tie together the cities and counties to provide an organization to develop local solar power. The main concern was if the six-county region would develop solar energy itself or have it "done to us and for us." The group formed a council of governments to enable the formation of the power authority. Another model is siting projects developed by IPPs, which obtain permits from local counties. This model involves decision-making by Xcel Energy and the PUC because the IPPs must have a PPA to guarantee sales of their electricity in order to obtain the financing necessary to build the project.

Finally, because of the large percentage of public lands in the SLV, some of the permits for large-scale solar projects will be granted by the BLM.

9.3. Proponent Statements in Light of Study Findings

Three statements by those advocating utility-scale solar development and renewable energy transmission lines seem to go to the heart of some of the concerns of the SLV community.

(1) *Utility Goal for Large-Scale Generation*
The ultimate intention of Xcel and Tri-State is to provide transmission for a very large amount of electricity from the SLV, as voiced by Karen Hyde in PUC hearings on the line: "The potential solar resources in the San Luis Valley are in the thousands of megawatts, not hundreds. Artificially limiting transmission to the bare minimum necessary to connect resources that result (only) from the 2007 Resource Planning Docket would not be wise."[86] This statement lends support to the concerns, valid or not, that CSP development would result in the "industrialization of the Valley floor." The utilities expect that, in the future, "thousands of megawatts" would be generated in the Valley, which would require tens of thousands of acres. If the expectation were not there, the transmission line would not have been proposed. Indeed, the goal of significant reductions in carbon-dioxide emissions would be enhanced by such large-scale development.

(2) *Statewide Permitting*
The WGA Task Force recommended statewide permitting for CSP siting; this approach appears also to be favored by GEO staff. This process could, if actualized, remove local control of land-use decisions on CSP facility siting and transmission siting. Such a decision would intensify concerns in the Valley.

(3) *Solar Exemption from Sales and Property Taxes*
The WGA Task Force report recommended that central solar plants be exempted from sales and property taxes, stating that the loss of these taxes would be offset by new tax revenues generated by economic activity caused by the plants.[87] If enacted, this proposal would remove one of the key benefits anticipated by county commissioners for their local counties—the property tax revenues to be paid by the facilities themselves. These tax revenues are viewed as fundamental benefits of solar development by SLV stakeholders. The prevention of property tax revenues for local counties could negatively affect perceived distributive justice; that is, the perceived equity of the distribution of costs and benefits from renewable energy development.

As they become known and understood by SLV residents, these policy recommendations could intensify community opposition to CSP siting and the proposed transmission line over La Veta Pass. They provide some factual support to fears about the loss of local control over land uses and about marring of scenic areas and viewsheds. The recommendations could also mean a potential loss of relative economic advantages to the tax base of SLV counties often cited by stakeholders as a benefit of CSP development.

9.4. Conclusions

Several broad conclusions emerge from the study as follows.

- *Widespread support for solar energy development*
  Although the preferences on how it should be done vary, a common element of SLV stakeholders is broad favorability toward the development of renewable energy resources, and a great deal of local pride in the SunEdison plant near Mosca and other solar facilities being developed.
- *Lack of knowledge and need for education*
  The SLV lacks knowledge on renewable energy. There is widespread confusion about the difference between PV and solar thermal electric systems and lack of knowledge about how CSP differs from other solar thermal electric systems.
- *Complex decision processes lack transparency*
  CSP and transmission decision processes are extremely complex. They currently involve a confusing mix of entities and procedures. There seems to be a tendency to keep relevant information secret, open only to those in the power structure for proprietary and adversarial reasons or "in the know" for political reasons. An integrated and transparent decision process that fully integrates federal and state policies with concerns of the local area would increase effectiveness and social acceptance of renewable energy siting decisions.
- *Public participation processes are more procedural than substantive*
  Public meetings and hearings have been held by IPPs, BLM, RUS, PUC and others at which supporters of the transmission line and opponents of it have testified. Although these events provide a formal mechanism for local-area stakeholder to be heard, the processes duplicate each other and are burdensome to the local community.
  John Villyard, Chief Executive of the SLV REC, said: "I wish NREL had never put the Valley on the solar map. Every week I've got developers coming into our offices with plans to put solar plants in the Valley. ...[The Valley] just feels overwhelmed...."[88] The effect of the public testimony on bureaucratic and private industry decisions is as yet unclear, especially in the milieu of the adversarial proceedings at the PUC regarding the transmission line, which could continue for some time.
  The outcome of these processes could affect perceived procedural justice; that is, the perceived inclusiveness of the decision process.
- *Social-psychological effects are not understood*
  Stakeholders within the SLV mentioned social-psychological benefits from solar development and also potential harms.
  Little attention has been paid to these types of impacts in project siting, yet they involve fundamental issues of community identity.

- *Organized opposition to the La Veta Pass transmission line*
  The advent of an organized opposition to the proposed transmission line over La Veta Pass that involves several organizations, including the Trinchera Ranch, is problematic for the successful resolution of this particular project.
- *Outside-the-Valley stakeholders are not fully informed about nor engaged in Valley concerns.*

The problems with CSP in the SLV reveal broader issues with respect to renewable energy siting, although CSP problems are not necessarily identical with problems in siting wind facilities.

For example, when wind turbines are sited, other land uses such as grazing or growing crops can continue, but when utility-scale solar facilities are installed, the land cannot be used in any other way. While also being land intensive, CSP precludes other simultaneous uses of the land.

9.5. State Policy Recommendations

A set of recommendations relative to state policy actions that could be taken is discussed. These are related to the report's findings.

- *Ensure that SLV community concerns are addressed and that Valley stakeholders and residents are engaged in CSP-related and transmission decisions.*
- *Develop education programs for SLV residents.* Local institutions of higher education should be funded to conduct education programs for SLV residents on energy efficiency, energy conservation, renewable energy distributed generation, and central-station systems. This funding should be provided by a consortium of partners, including developers, state government, and federal government agencies.
- *Integrate local decision-makers.* A San Luis Valley energy-water-land group should be formed to represent Valley-wide energy, water, and land-use concerns in integrated siting and other decision-making processes at regional, state, and federal levels. This group should conduct analyses and take informed positions—reaching consensus as often as possible—on the integration of broad local and regional energy, land-use, and water concerns. Its primary purpose would be to integrate natural resources expertise. The activities of this group should be funded by federal, state and local programs, environmental organizations, and the renewable energy industry. Cognizant federal agencies could fund the organization and support of this group, the administration of which could be based at an SLV institution of higher education.[89]
- *Offer workforce training.* Local colleges should be funded to conduct workforce training on vocations related to employment opportunities in connection with distributed and large-scale solar development in the Valley in partnership with the state community colleges and state and local workforce development officers.
- *Provide training for SLV decision makers.* Local educational organizations, in partnership with the legal educators, should organize workshops and seminars for

city, county, and federal officials in the SLV on the statutory and regulatory implications of renewable energy decision-making at the state and federal levels.

- *Proactively approach spiritual interests.* The spiritual aspects of solar development should be explored as soon as possible.
- *Evaluate models of taxation.* Alternative models of taxation that would compensate the SLV community for the expenses involved in mitigating the impacts of CSP and transmission development should be investigated.
- *Potentially form landholder energy associations.* Such associations could create a public sphere for understanding and developing policy positions supported by the citizens of Colorado in renewable energy resource areas, beginning with the SLV.

9.6. Concluding Remarks

The SLV could become a laboratory for new approaches to siting decision-making, one that is sorely needed at the national level as other controversies concerning renewable energy and transmission siting at Cape Cod, San Diego, Medicine Lake, and other areas, make clear. As a human community, not an unpopulated desert, the SLV and its scenic and wildlife corridors, federal lands, and National Parks must be respected and protected. The opportunity to protect vulnerable communities and to engage them in CSP and transmission decisions, benefiting from their expertise rather than treating them as adversaries, should parallel in significance the investment in solar technology.

No matter what the ultimate outcome is in the SLV, the development of the SLV Power Authority should be further investigated in that it could be a useful test case on how small- to medium-scale solar installations integrated with utility substations and smart grids could be organized and managed.

The most effective, socially acceptable institutional arrangements for generation and distribution of solar power may yet need to be invented.

Afterword

On June 3, 2010, the *New York Times*[90] reported on the SLV controversy, terming it a "bitter fight over solar's future," and pointing out that the path to a new energy future "will not be without its own messy entanglements of politics, power, and place." The Times story reported that in May 2010, Xcel Energy said that delays in the project, which they blamed on Louis Bacon and his lawyers, had "undermined the potential development of solar power in Alamosa."

A utility spokesman reportedly said that the utility "may have to go somewhere else." However, a spokesman for Mr. Bacon said he believed the utility was reconsidering for economic reasons and scapegoating Mr. Bacon.

"Blame the billionaire," he said. On June 5, 2010, the *Denver Post*[91] reported that Xcel Energy had asked the Colorado PUC to reduce the amount of solar power that it is required to obtain.

Xcel Energy reportedly said it could modify current transmission lines and provide 185 MW of solar energy by a 2015 deadline, a 48% reduction in existing requirements for solar power.

APPENDIX A. SUMMARY OF THE BLM PEIS PROCESS: THE US BUREAU OF LAND MANAGEMENT'S AND US DEPARTMENT OF ENERGY'S PROGRAMMATIC ENVIRONMENTAL IMPACT STATEMENT FOR UTILITY-SCALE SOLAR ENERGY DEVELOPMENT

A-1.1. Background

As part of its ongoing efforts to increase domestic energy production and ensure greater energy security, the Bureau of Land Management (BLM) initiated a joint Programmatic Environmental Impact Statement (PEIS) with the Department of Energy (DOE) to assess the environmental, social, and economic impacts associated with utility-scale solar energy development on BLMmanaged public land in six western States: Arizona, California, Colorado, Nevada, New Mexico, and Utah. The Notice of Intent (NOI) was published in the Federal Register on May 29, 2008.[92]

A-1.2. Chronology of Key Events

As the federal government moved to implement a policy of increased development of solar resources, it became clear that the public lands would have to be used for this purpose. The BLM had a backlog of applications for permits to develop large-scale solar facilities on public lands. However, permits cannot be issued for any activity that is not included in the Bureau's Resource Management Plans (RMPs); therefore, the RMPs have to be modified to allow permits to be granted. To facilitate this process, the BLM and DOE joined together to conduct a programmatic environmental impact statement on large-scale solar development in six Western states. Once the PEIS process is completed, it would be for permits to be granted more readily than otherwise would have been the case. Table A-1 summarizes the key events in the PEIS process and the designation of solar study areas in the six states.

The BLM has conducted two different PEISs. One was for the transmission corridors, which has been completed and the record of decision announced. The other, which is ongoing, was for solar development on public lands. This PEIS is the focus of this section of the report.

A-1.2.1. PEIS on Energy Transmission Corridors in the West

The joint PEIS, released at the end of 2008, identifies energy corridors in the West for transmission and distribution lines that will help facilitate the development of renewable energy resources. The energy corridors could also carry pipelines for oil, natural gas, and hydrogen. Approximately 5,000 miles (8,000 km) of energy corridors are located on BLM-managed lands, while nearly 1,000 miles (1,600 km) of energy corridors are on US Forest

Service lands. Roughly 120 miles (190 km) of corridor segments are on lands managed by the Bureau of Reclamation, the National Park Service, and the Department of Defense.[93]

The BLM and the *US Forest Service* issued *Records of Decision* in mid-January 2009 to amend 130 of their land use plans to support the designation of more than 6,000 miles (9,700 km) of *energy transport corridors* on federal lands in 11 *Western states*. The amendments were based on analyses presented in a Final *PEIS* that was prepared by the BLM, DOE and the US Departments of Agriculture and Defense as part of their work to implement the *Energy Policy Act of 2005*.

Table A-1. Chronology of Key Events in the Management of BLM Public Lands for Large-Scale Solar Energy Development

Date	Key Event
2008 – ongoing	BLM has a backlog of hundreds of solar energy and wind project applications for siting permits on public lands
December 2008	BLM issues a Final PEIS for itself, DOE, USDA, and USDOD identifying energy transport corridors to facilitate the development of renewable energy resources
January 2009	Following on the energy corridors PEIS, BLM and USFS issued Records of Decision amending 130 land-use plans to designate 6,000+ miles of energy corridors in 11 Western states
May 28, 2008	BLM-DOE initiated another joint PEIS for utility-scale solar energy development on BLM-managed public lands in six Western states: Arizona, California, Colorado, New Mexico, Nevada, Utah with Notice of Intent (NOI) published in Federal Register as of this date.
May-July 2008	Public Scoping Comments period on the joint PEIS
March 11, 2009	Secretary of Interior Ken Salazar issues Secretarial Order No. 3285 announcing a policy goal of identifying and prioritizing specific locations best suited for large-scale production of solar energy
May 5, 2009	Secretary of Interior Ken Salazar announces formation of four Renewable Energy Coordination offices in Arizona, California, Nevada, and Wyoming and smaller renewable energy teams in Colorado, Nevada, New Mexico, and Oregon
June 30, 2009	BLM designates 21,000 acres in the San Luis Valley of Colorado as Solar Energy Study Areas (SESAs) (of 670,000 acres total) with NOI published in the Federal Register.
July-September 2009	BLM-DOE consider public comments on the PEIS for solar development.
September 14, 2009	Last day of public scoping meetings on the PEIS
Summer/Fall 2010	Planned release of Draft PEIS
Late fall 2010 (TBD)	Final PEIS and Record of Decision to be released

A-1.2.2. Related Federal Actions

On May 5, 2009, Ken Salazar, Secretary of the Interior, pledged to create four Renewable Energy Coordination Offices, one each in California, Nevada, Wyoming and Arizona. In addition, Salazar proposed the creation of smaller renewable energy teams located in New Mexico, Idaho, Utah, Colorado and Oregon. Such offices will expedite production of renewable energy on public lands.[94]

The renewable energy offices and teams are being set up to expedite applications, processing, reviews, and permitting of renewable energy projects. Salazar said the Interior Department is investing $41 million through the President's economic recovery plan to facilitate a rapid and responsible move to large-scale production of renewables on Bureau of Land Management lands. Currently, the Interior Department has a backlog of hundreds of solar energy and wind project applications for siting on BLM lands. The new offices will also improve the BLM's coordination with state agencies and other federal agencies, including DOE and the *US Environmental Protection Agency.* [95]

A-1.2.3. PEIS on Solar Energy Development

On May 28, 2008, BLM and DOE announced their new PEIS process. With Notice of Intent (NOI) published in the Federal Register on June 30, 2009, the BLM announced the availability of maps that identified 24 tracts of BLM-administered land for in-depth study for solar development, and requested public comment on those study areas. Public comments were accepted in the six Western states from July-September 2009.

According to the *Solar Energy Development Programmatic EIS* website, the release of the Draft Solar Energy Development PEIS was originally scheduled for spring of 2009.

However, BLM and DOE have agreed to postpone completion of the Draft PEIS in order to more closely align the proposed alternatives in the Draft EIS with Secretarial Order No. 3285 (issued March 11, 2009 by the Secretary of the Interior), which announced a policy goal of identifying and prioritizing specific locations best suited for large-scale production of solar energy.

A-1.2.4. Designation of the Solar Energy Study Areas (SESAs) in the San Luis Valley

With anthropogenic climate change acknowledged by the government, industry, and the broader public coupled with the high cost of oil and gas, national focus on renewable energy targets and appropriate technology is growing.

As noted earlier, research on solar energy potential indicates that the San Luis Valley (SLV) is a prime location for solar development, making it one of the most favorable locations in Colorado to develop solar projects, especially utility-scale solar energy projects.

The BLM's plan for *SESAs* totals 670,000 acres, including 21,000 acres in Colorado's San Luis Valley. The goal, according to the BLM, is to speed project permitting for utility-scale solar power projects.

The 24 tracts being considered for the study areas are being evaluated for their environmental and resource suitability for large concentrating solar power and photovoltaic systems installations.[96]

In the course of the PEIS analyses, the agencies have identified a number of tracts of BLM-administered land for in-depth study for solar development. Four study areas have been identified in Colorado: Antonito Southeast (9,598 acres), De Tilla Gulch (1,522 acres), Fourmile East (3,882 acres), and Los Mogotes East (5,909 acres). [97]

The Colorado locations could generate up to 4,182 MW of solar power if fully developed, according to the US Department of the Interior. If areas are deemed suitable for large solar projects, then companies would have the option of building installations with a capacity of 10- MW or more.

According to the Secretary of Interior, companies proposing projects in these areas would be eligible for faster permit processing.[98]

Because developing these types of facilities has environmental implications, public discourse is encouraged, and is, in fact, required through NEPA.

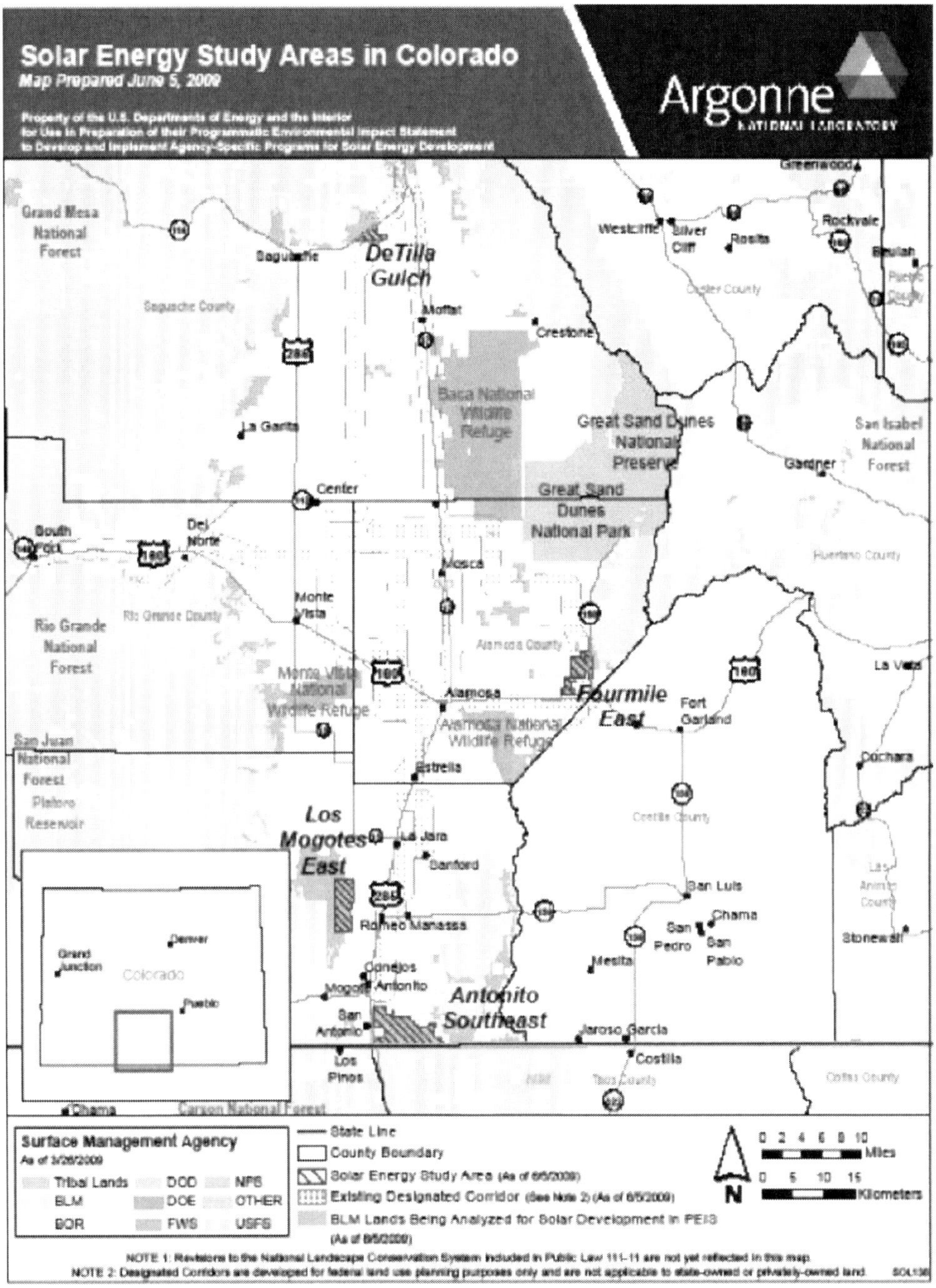

Figure 1. Solar Energy Study Areas in Colorado (June 5, 2009) Credit: Argonne National Laboratory.

A-1.2.5. Public Comment Process

The BLM accepted comments on the scope of the PEIS through July 15, 2009 and held public scoping meetings in the six states through September 14, 2009.[99] The *Solar Energy Development Programmatic EIS* website is the online center for public information and contains 35 public comments from organizations and individuals recorded through September 14, 2009. The organizations and individuals represented include:

Organizations:
Arizona Wilderness Coalition
Californians for Western Wilderness
California Native Plant Society
California Wilderness Coalition Center for Native Ecosystems (Pollock, Robertson)
Colorado Environmental Coalition (Curtiss)
Defenders of Wildlife
Desert Protective Council
Grande Water Conservation District
Great Old Broads for Wilderness (Egan)
Mesa County, Planning and Economic Development (Larsen, Price)
Mojave Desert Land Trust
National Trust for Historic Preservation (Pahl, Poston, Hartig, Schwendler)
National Wildlife Federation
Natural Resources Defense Council
New Mexico Wilderness Alliance
Rio Grande Water Conservation District (Vandiver)
Saguache County Board of County Commissioners (Pace, Spearman, Joseph)
San Luis Valley Ecosystem Council (Canaly)
San Luis Valley Water Protection Coalition (Crowley)
Sierra Club
Soda Mountain Wilderness Council
Southern Utah
The Wilderness Society (Daue)
Western Resource Advocates (Darin)
Western Resource Advocates
Wilderness Alliance
Wild Utah Project
Wyoming Outdoor Council

Individuals:
Austin, Gay (Gunnison; CO)
Barker, Claire (2) (Mosca, CO)
Barker, Donald (CO)
DeHerra, Guillermo (Location Withheld)
Diss, Orville, (Center, CO)
Howard, Eugene D. (Denver, CO)

Mitson, Loretta M. (Manassa, CO)
Naylor, Thomas A. (Fairplay, CO)
Plutschuck, Donna M. (Lakewood, CO)
Smith, Jeff (Location Withheld)
Smith, Cecelia (Crestone, CO)
Zenoit, Michael (Denver, CO)

Although it was not possible to include all public comments posted on the PEIS website the selection of comments included in this report represents a range of issues and their pros and cons. The comments fall under four broad categories, defined as (1) energy systems impacts (renewable/conventional), (2) environmental impacts, (3) economic impacts, and (4) the PEIS process itself.

Energy Systems (Renewable/Conventional)

Issues include renewable development, dependency on fossil fuels, use of conventional and other (i.e. wind energy sources), GHG emissions, job creation and cost benefit analysis. Related submission(s) are presented below.

Private citizen Mitsin commented in part: "*I believe that solar development should be treated just like agriculture. We need to support large scale agribusiness as well as small scale family, community (GSA) farms. We need to encourage large scale solar development as well as small-scale solar generation on every possible rooftop and home. BUT, we need to be very careful where we site large-scale agribusiness as well as where we site large scale solar energy generators. Utilizing public lands for solar generation is not a good choice.*

As an individual who utilizes passive solar collectors on my home, I do not have to be convinced of the viability of small-scale solar utilization. I do not believe that the power distribution companies have really explored the possibility and necessity of generating power closer to its point of use. Transporting power over long distances only serves to reduce the net gain of kWh. Ideally, solar generators need to be sited closer to the cities, where the production can be maximized." Adding: " If the Solar Rewards Program was adequately funded instead with the millions of dollars that are earmarked to be squandered on hundreds of miles of high voltage lines, and the cost of acquiring lands for right-of-way, maybe we could get more of the population behind a real grassroots interest in satisfying and generating more of our energy needs right in our own communities. Let's not 'pave paradise and put up a parking lot.' There is plenty of private land to use for solar generation. Don't use our public lands for that, or we will never get them back. Public lands are sacred." [100]

The National Trust for Historic Preservation commented in part: "*The National Trust for Historic Preservation (National Trust) commends the efforts of the Bureau of Land Management for (BLM) to identify appropriate areas for solar energy development while limiting impacts to significant cultural resource. We feel strongly that by incorporating the results of tribal consultation and previous cultural resources inventories into the development of the solar Programmatic Environmental Impact Statement (PEIS) and the specific Solar Energy Study Areas (SESAs), BLM can facilitate efficient and cost-effective renewable energy development while protecting the invaluable, significant cultural resources that are present on America's federal public lands. At the same time, we believe that efficiency and reduction of energy use at private and public scales should be a focus of rent and future energy* *planning. Rather than simply producing and transmitting more energy, we would simultaneously work to reduce our energy needs."*[101]

Environmental Impacts

Issues include concern for ecosystems, species analysis, and distribution. Related submission(s) are included below.

San Luis Valley Ecosystem Council (SLVEC) comments in part: *"The mission of SLVEC is to protect and restore—through research, education, and advocacy— the biological diversity, ecosystems, and natural resources of the Upper Rio Grande bioregion, balancing ecological values and human needs. SLVEC works as the only local public lands advocacy organization that is concerned about protecting and restoring intact ecosystems and wildlife corridors, from the mountain peaks to the rivers along the valley floor, and into New Mexico. We encourage both a national and a regional conversation on energy use and, especially, on fossil fuels and their impacts to climate change. It is imperative that our country makes the transition to the use of renewable energy sources. The warming effects are being felt in the San Luis Valley, as in other parts of the world, and are impacting wildlife, water supplies, and forest health. We believe that renewable energy can offer a clean, affordable, sustainable, and environmentally friendly alternative. We support a measured approach, however, to the switch to alternatives. We recognize the unique and valuable aspects of the San Luis Valley. We understand that the Valley has enormous potential in the area of solar production, and has a long history of supporting solar energy on a smaller scale. We encourage the development of renewable energy strategies that will promote the long-term health and well being of the community, and protect the environment, critical habitat, wildlife, sensitive corridors, and water, as well as the history and culture of this agro-pastoral community. We urge the DOE and BLM to take a long term view when considering the scale, siting, water demands and the building of new transmission lines that will be required to accommodate Utility Scale Solar development in a culturally and ecologically sensitive area like the San Luis Valley (SLV).*

It is imperative that solar development remain responsible and that renewable energy development does not compromise this area's unique values. *We recommend a national model of appropriate energy development based on what is currently being implemented in European countries. They appear to exercise a threefold strategy; emphasis on flexibility in size and scale fitted to location and need, constructing open ended systems that can rapidly integrate new technologies, and suitably subsidizing research and development that encompasses a range of alternative energy sources.*[102]

Rio Grande Water Conservation District commented in part: *"The District Board of Directors does not have a policy of opposition to solar power. However, we are very aware that some types of solar generation can use significant amounts of water. In fact, we are lead to believe that some solar generation techniques use almost the same amount of water as a coal-fired generation system. As you know, the State of Colorado is a prior appropriation state. The Federal Government and its agencies have been joined in the State water rights process pursuant to the McCarran Amendment. And, pursuant to that joinder, the Bureau of Land Management has sought and obtained any reserved water rights to which it is entitled. Any water required for solar generation must therefore be acquired pursuant to Colorado law and will be governed by the doctrine of prior appropriation."*[103]

Private citizen Austen wrote: "*Dear Personnel, I am writing to you about my concern regarding the Solar Energy Development project proposed on BLM lands across the West. I am concerned that this project may damage or destroy threatened, endangered, or sensitive plant and/or rare lichen populations. Are plant surveys, plant BA's, and plant BE's going to be conducted by qualified Botanists for the project area?* [104]

Private citizen Barker commented in part: "*The economic benefit to communities affected—specifically individuals whose property value plunges and community business economics-dependent on non-industrialized ambiance—are not well regulated or methodically reimbursed for losses incurred secondary to solar and transmission development. Much falls under 'eminent domain.' All industry-solar included-require some water-if only to wash panels free of dust-which is plentiful in many of the scoped regions. Other technologies to provide storage capability of energy-utilize the equivalent in water consumption to present agribusiness-BUT-utilizes it 12 months out of the year-instead of just three. Unregulated or legal changes in regulations of water use-make sunny, arid climates very vulnerable to continuing water issues that devastate the local eco-systems. The agricultural industry is already struggling with this issue, another "layer" of industrial usage permits would be paradoxical under the guise of "green". At present-I acknowledge that the energy industry/energy consumer is at a crossroads-under the time limit of legal change in Colorado-to increase "green energy consumption and development by 10%. We are also at a crossroads where moving too fast and in the "path of least resistance and cost" will cost our future generations dearly. We KNOW this-even though the technology is new-the pitfalls are predictable. Let us NOT repeat history-please. Uphold public lands for public enjoyment, open space, environmental sanctity and eco-system balancing. The issue is already difficult enough without adding another layer of potential problems.*"[105]

Economic Impacts

Issues include the benefits of job creation, concern over property values, tourism due to a changed landscape. Related submission(s) are included below.

Private citizen Diss commented: "*I would encourage use of the BLM land in the San Luis Valley in Colorado for Solar development. The SLV is one of the most depressed areas in the entire country. The entire economy of this valley is based on the price of potatoes. With increasing pressure from State and other entities to cut back on water consumption, this valley will literally dry up and blow away without some other economic base that doesn't require much water.*"[106]

Private citizen Smith commented: "*My thought is that we develop our natural resource (oil, gas and nuclear) before we subsidize renewables. The amount we can generate from renewables is minuscule to that of other sources. Not to mention the job loss in developing renewables is nearly 3 to 1 vs. developing our natural resources. A great example is the solar panels put atop the museum of science and natural history in Denver. It was considered by the Administrators to put solar panels on their building but after discovering the return on investment took 114 years and the panel's life was 25 years it logically was shelved. Then appears the government who is willing to take taxpayer money to fund renewable energy. Not with logic but with taxpayer money to appease the greenies. What a waste and if that kind of investment continues the taxpayers will run out of money. As a business decision and that's how it has to be looked at it's a no brainer.*

Private citizen Howard commented: "*Please do proceed with evaluating public lands for solar. We need to break our country's dependence on OIL and instead of DRILLING in sensitive areas that could spoil the land and doesn't move to solve the problem, install SOLAR PV, THERMAL and Wind on public lands that go to make our country a safer place and energy independent from the rest of the world. THIS is a GREAT USE of Public lands that belong to all Americans. Thank you for listening to me.*"[107]

Private citizen Plutshuck commented: "*Finally! The BLM & the US Government have to help in whatever way possible with developing renewable energy. Not only does it put people to work, it maintains Americans' standard of living while reducing greenhouse gases. Please, install solar and wind power.*"[108]

Private anonymous citizen commented: "*I support the choice of the solar study areas in Colorado. I live in Antonito and not too far from the study just south of Antonito. This location is ideal for the purpose intended and solar energy is vital to the future of our country. I hope that the government will help those of us in small rural towns to obtain affordable means of using alternative energy sources such as solar and wind power. I also think that wind power study areas should be established; and, the locations chosen for the solar study are also viable for wind power study.*"[109]

PEIS Process

Issues include regulatory practice, water and land use and appropriation, siting, and transmission, proximity to urban centers, infrastructure, robust planning and assessment. Related submission(s) are included below.

The Wilderness Society joint submission stated in its introduction: "*We support BLM's to develop clear and comprehensive guidelines for responsible solar energy development, identify lands appropriate for solar projects as open for development, and close all other lands to development as part of the Solar Energy PEIS. The release of proposed SESAs for public review and comment is an important next step showing the BLM's commitment to this approach and providing more detail on how it can be accomplished. We are encouraged by the BLM's statements that important screening criteria (including critical wildlife habitat, special management areas, and visual resources) have already been applied to SESAs. Further, establishing SESAs better enables a landscape-level analysis of solar development and associated transmission on public lands in the West.*"[110]

Mesa County Department of Planning and Economic Development commented in part: "*The PDF maps on the Solar PEIS website provide insufficient detail to evaluate the 4 sites proposed in Colorado. The data used to create these maps should be made available for download in formats that can be used by popular GIS software. Additionally, maps should be made available interactively online similar to http://www.nrel.gov/eis/imby/.*" "*The PEIS should take into account existing infrastructure and proximity to urban areas in a formula for calculating suitability of BLM land for solar development. The map titled "Solar Potential from Concentrating Collector," shows that the tracts in southern Colorado that are part of this PEIS provide 6.0 to 6.5 kWh/sq. meter/day. Other areas managed by BLM of slightly less available solar energy with values of 5.5 to 6.0 kWh/sq.meter/day that are closer to urban areas and are located along existing transmission lines were not considered for this PEIS. Electrical transmission loss over long distances may negate the slightly higher (10%) solar energy available in the PEIS study area. New electrical transmission corridors and higher capacity transmission lines will consume land and create their own negative environmental* *and aesthetic impacts. As part of the draft Solar PEIS, areas slightly less than 6.0 to 6.5 kWh/sq. meter/day, should be considered and included as one of the alternatives if they are more efficient by their close proximity to existing infrastructure."*[111]

Saguache County Board of County Commissioners commented in part: *"The solar energy generation advances at a rapid pace. Yet, the new large-scale fields being developed are new to all of us. Little is known of the potential impacts of the infrastructure and processing of solar energy at this scope. It is essential that adequate baseline information be available, and monitoring reviews be included in arrangements between the BLM and industry, to assure early detection and mitigation of currently unforeseen impacts."* [112]

Private citizen Barker commented in part: *"Solar development—while 'greener' in SOME ways, on a utility-scale basis— disrupts the landscape, denies access, develops access where none was and changes the shade/sun/water cycle so critical to many of the areas being "scoped". Thousands of solar panels precede the problem of removal of obsolete panels-which are toxic-and presently non-recyclable. I fear greatly the setting of "precedence" through "NEED" under the guise of green energy. Denying known problems of environmental/socio-economic damage to communities affected and warping industrial sized solar development potential to fit a gas/coal model. Solar is most effective close to its source of consumption. Utility scale or personal panel placement. ALL of Colorado (Western United States) has great solar potential and tracts of industrial land, private land and public right of ways perfect for locale community use-specific development. Hurrying to seek a path of least resistance, under the pressure of federal and state scrutiny, will make it more difficult or impossible in the future, to mitigate applications for development of any and every kind on our gorgeous public lands. Utility scale solar energy development presently is tied into transmission line development. Connecting large tracts of land, over miles of more land to reach each other. Toxic views, migratory and native bird threats, extension of development in even more "open space" beyond the immediate utility solar plant, and only token legal attention to environmental/socio-economic impacts of the communities and ecosystems involved."* [113]

Private citizen DeHerra commented in part: *"In the American West there are old sayings that in simple ways tell the story. For Solar Energy Project Development in the San Luis Valley there is such an appropriate old saying, 'you have to strike when the iron is hot,' which is a must for branding calves with a hot iron. In the United States today there is tremendous interest in renewable energy and a multitude of government incentives for such projects. The heightened interest by the American people and the public policy and incentives created by the new Obama administration are driven by the global financial crisis, price of oil nearing peak supply, turmoil in the Middle East, and the need to move America to new energy sources and away from foreign oil. However there are limitations to the market opportunities, so as the market for solar energy heats up, it would be smart to 'strike when the iron is hot'. What this means for the Bureau of Land Management (BLM) in the rule making process of this programmatic EIS, is that action taken to allow the private sector to utilize these public lands for solar energy development must be swift, with minimal or reasonable regulation adopted in an open process to allow for fair competition. In addition, other measures by federal and state governments will need to be initiated and/or continued for solar energy to be competitive with fossil fuels for commercial power generation." From page 7 - "The PEIS effort should continue and rules and regulations issued to allow* *companies to compete for fast track licensing of the two tracts of land. This new technology and the tremendous human and natural energy potential have certainly motivated me to get involved and likely return to my homeland. Over the last half century, one of the greatest exports from the San Luis Valley has been people; it is now time to bring them home as new jobs and opportunities are created."*[114]

A-1.3. Summary

The PEIS designated tracts of US public lands in the West as prime zones for utility-scale solar energy development, funded environmental studies, opened new solar energy permitting offices and is working to speed reviews of industry proposals. Under this initiative, 24 tracts of Bureau of Land Management-administered land located in six western states, known as Solar Energy Study Areas, would be fully evaluated for their environmental and resource suitability for large-scale solar energy production. The objective is to provide landscape-scale planning and zoning for solar projects on BLM lands in the West, allowing a more efficient process of solar development. The schedule for release of the draft PEIS will be determined after the evaluation of comments from the current scoping period addressing the solar energy study areas. The Draft PEIS is expected to be available for public review and comment in late summer or early fall 2010. [115]

APPENDIX B. CHRONOLOGY OF KEY EVENTS IN THE SLV TRANSMISSION SITING CONTROVERSY

Table B-1. Chronology of Key Events in the Controversy over Siting New Transmission Lines and Substation in the San Luis Valley

Date	Key Event
March 2007	S.B. 100 passes in the Colorado legislature and is signed into law by Governor Ritter
January 2009	BLM and USFS issued Records of Decision amending 130 land-use plans to designate 6,000+ miles of energy corridors in 11 Western states
April 2009	H.B. 1300 authorizes the "Clean Energy Development Authority" (CEDA), a Colorado "transmission authority".
May 2009	Applicants Xcel Energy and Tri-State Electric Generation and Transmission Association file for a Certificate of Public Convenience and Necessity (CPCN) (Colorado PUC Docket Number PUC Combined Docket #08E324 and 325)
June 2009	Tri-State submits "Alternative Evaluation" to the USDA which evaluates alternative transmission routes
October 28, 2009	Interveners in Combined Docket #08E324 and 325 submit answer testimony
November 9 and 10, 2009	The PUC holds public hearings in Walsenburg and Alamosa, Colorado, concerning the application for a CPCN by Xcel Energy and Tri-State
December 2, 2009	Interveners in Combined Docket #08E324 and 325 submit rebuttal and answer testimony
January 18, 2010	Sur-rebuttal Testimony was filed by applicants, WRA and Trinchera Ranch

REFERENCES

Alvarez-Farizo, B. & Nick Hanley. (2002). Using conjoint analysis to quantify public preferences over the environmental impacts of wind farms. An example from Spain. *Energy Policy, 30*(2), 107-116.

Babiuch, William M. and Barbara C. Farhar. 1994. *Stakeholder Analysis Methodologies Resource Book*. NREL/TP-461-5857, 104 pp.

Beierle, T.C., & J. Cayford. (2002). *Democracy in practice: Public participation in environmental decisions*. Washington, D.C.: Resources for the Future.

Bell, D., T. Gray, & C. Haggett. (2005). The "social gap" in wind farming and decisions: Explanations and policy responses. *Environmental Politics, 14*, 460-477.

Carlman, I. (1982). Wind energy potential in Sweden: The importance of non-technical factors. *Fourth International Symposium on Wind Energy Systems*. Stockholm, 335-348.

Colorado Governor's Energy Office. (2008). Connecting *Colorado's renewable resources to the markets*. Report of the Colorado Senate Bill 07-091, Renewable Resource Generation Development Areas Task Force, revised edition. Denver: Governor's Energy Office.

Colorado Governor's Energy Office. (2009). Renewable *energy development infrastructure: Connecting Colorado's renewable resources to the markets in a carbon-constrained electricity sector*. The REDI Report. Denver: Governor's Energy Office.

Devine-Wright, P. (2005). *Beyond NIMBYism: Towards an integrated framework for understanding public perceptions of wind energy*. Wind Energy, 8(2), 125-139.

Devlin, E. (2005). Factors affecting public acceptance of wind turbines in Sweden. *Wind Engineering, 29*, 503-511.

Eiser, J. R., R. Spears, & P. Webley. (1988). Predicting attitudes to oil and to nuclear energy. *Journal of Environmental Psychology, 8*(2), 141-147.

Ellis, G. (2004). Discourses of objection: Towards an understanding of third-party rights in planning. *Environment and Planning, 36*, 1549-1570.

Eltham, D.C., G.P. Harrison, & S.J. Allen. (2008). Change in public attitudes towards a Cornish wind farm: Implications for planning. *Energy Policy, 36*, 23-33.

Enzensberger, N., M. Wietschel, & O. Rentz. (2002). Policy instruments fostering wind energy projects – a multi-perspective evaluation approach. *Energy Policy, 30*, 793-801.

Goodnough, A. (2010). For Cape Cod wind farm, new hurdle is spiritual. *The New York Times*. January 4. Retrieved March 21, 2010, from http:// nytimes.com/2010/01/05/ science/earth/05wind.html.

Goodnough, A. (2009). Wind farm off Cape Cod clears hurdle. *The New York Times*. January 16. Retrieved March 221, 2010, from http://nytimes. com/ 2009/01/17/us/17wind.html.

Guerrerio, J. (2009). The boom-bust cycles of energy-rich Western Colorado. *The Energy Examiner*. Retrieved February 9, 2009, from http:// wilderness.org/content/boom-or-bust-plowing-pace-oil-and-gas-drilling.

Heiman, M.K. & B.D. Solomon. (2004). Power to the people: Electric utility restructuring and the commitment to renewable energy. *Annals of the Association of American Geographers, 94*(1), 94-116(23).

Jaffe, M. (2009). Drawing the powerline, high voltage transmission plans in San Luis Valley spark a battle over renewable-energy development. *The Denver Post*, December 13, 1K, 8K.

Johansson, M., & T. Laike. (2007). Intention to respond to local wind turbines: The role of attitudes and visual perception. *Wind Energy, 10*, 435-451.

Johnson, K. (2010). "Going Solar is Harder than It Looks, A Valley Finds," *The New York Times*. June 3, 2010.

Kahn, R. (2000). Siting struggles: the unique challenge of permitting renewable energy power plants, Electricity Journal 13:21-33. Juillerat, L. & K-Falls Herald & News. (2006). Dispute swirls around geothermal project. January 14. Retrieved on March 21, 2010, from Klamath Forest Alliance Web site: http://klamathforestalliance.org/ Newsarticles /newsarticle 20060115. html.

Lehr, U., J. Nitsch, M. Kratzat, C. Lutz, & D. Edler. (2008). Renewable energy and employment in Germany. *Energy Policy, 36*, 108-117.

Mehos, M. & B. Owens. (2005). *Siting utility-scale concentrating solar power projects*.Presented at the 2004 DOE Solar Energy Technologies Program Review Meeting, Denver, CO. NREL/CP-550-37086.

Migoya, D. "Xcel seeks relaxed solar requirement." *The Denver Post*. June 5, 2010, p. 6B.

Mills, D. & R. Morgan. (n.d.). Solar thermal power as the plausible basis of grid supply. Palo Alto, CA: Ausra, Inc.

Nadai, A. (2007). “Planning”, “siting” and the local acceptance of wind power: Some lessons from the French case. *Energy Policy, 35*, 2715-2726.

Painuly, J.P. (2001). Barriers to renewable energy penetration; a framework for analysis.” *Renewable Energy, 24*, 73-89.

Renewable Energy Development Infrastructure. (2009). Connecting Colorado's renewable resources to the markets in a carbon-constrained electricity sector. A Report of the Colorado Governor's Energy Office, December.

San Luis Valley Development Resources Group. (2007). *Comprehensive economic development strategy*, *1*, Sections I-IV, Analysis A-G.

San Luis Valley Development Resources Group. (2007). *Comprehensive economic development strategy*, *2*, Analysis H-P.

San Luis Valley Development Resources Group. (2007). *Comprehensive economic development strategy*, *3*, Analysis Q-Z, Sections V-XII.

Smith, D. (2010). Medicine Lake geothermal leases questioned in appeal. *Daily News*. March 16. Retrieved on March 21, 2010, from http://www.siskiyoudaily.com/news/x673421766/ Medicine-Lake-geother mal-leasesquestioned-in-appeal.

Smith, E.R.A.N., & M. Marquez. (2000). The other side of NIMBY Syndrome. *Society & Natural Resources, 13*, 273-280.

Stoddard, L., J. Abiecunas, & R. O’Connell. (2006). Economics, *energy, and environmental benefits of community solar power in California*. (NREL Subcontract Report NREL/SR-500-39291). Overland Park, KS.

Szarka, J. (2006). Wind power, policy learning and paradigm change. *Energy Policy, 34*, 3041- 3048.

Task Force on Renewable Resource Generation Development Areas. (2008). *Connecting Colorado’s renewable resources to the market*. Denver, CO: Governor’s Energy Office. Submitted to Colorado Governor Bill Ritter, Jr. and the General Assembly of the State of Colorado December 21, 2007, Revised July 2008. Report on the Colorado Senate Bill 07-091.

United States Court of Appeals, Ninth Circuit. 2006. Pit River Tribe; Native Coalition for Medicine Lake Highlands Defense; Mount Shasta Bioregional Ecology Center, Plaintiffs-appellants, v. United States Forest Service; Advisory Council on Historic Preservation; United States Bureau of Land Management; Calpine Corporation, Defendants-appellees. (No. 469 F.3d 768). Retrieved March 21, 2010, from *http://cases.justia.com/us-court-of-appeals/F3/469/768/622353/.*

US Department of Energy, Office of Energy Efficiency and Renewable Energy. (2007). Report to Congress on assessment of potential impact of community solar power for electricitygeneration (EPACT 2005 – Section 934, Department of Energy/GO-102007-2400). Toke, D. (2005). Explaining wind power planning outcomes: Some findings from a study in England and Wales. *Energy Policy, 33*, 1527-1539.

van der Horst, D. (2007). NIMBY or not? Exploring the relevance of location and the politics of voiced opinions in renewable energy siting controversies. *Energy Policy, 35*, 2705-2714.

Webber, M.E. (2008). Catch-22: Water vs. energy. *Scientific American Earth 3.0.* October, 34- 41.

Western Governors' Association. (2006). Clean and diversified Energy Initiative Solar Task Force report. January.

Wolsink, M. (1996). Dutch wind power policy: Stagnating implementation of renewables. *Energy Policy, 24*, 1079-1088.

Wolsink, M. (2000). Wind power and the NIMBY-myth: Institutional capacity and the limited significance of public support. *Renewable Energy, 21*, 49-64.

Wolsink, M. (2007). Planning of renewable schemes: Deliberative and fair decision-making on landscape issues instead of reproachful accusations of non-cooperation. *Energy Policy, 35*, 2692-2704.

Wustenhagen, R., M. Wolsink, & M.J. Burer. (2007). Social acceptance of renewable energy innovation: An introduction to the concept. *Energy Policy, 35*, 2683-2691.

Zweibel, K., J. Mason, & V. Fthenakis. (2008). Solar grand plan. *Scientific American, 298*, 64- 73.

End Notes

[1] Carlman 1982; Wustenhagen et al. 2007.
[2] Alvarez-Farizo and Hanley 2002; Devine Wright 2005; Devlin 2005.
[3] Wustenhagen, et al. 2007: 2685.
[4] Ibid.
[5] Eiser, et al. 1988; Heiman and Solomon 2004; Wolsink 2000.
[6] Bell et al. 2005; Kahn 2000.
[7] Wolsink 2007:2699.
[3] Wustenhagen, et al. 2007: 2685
[4] Ibid.
[5] Eiser, et al. 1988; Heiman and Solomon 2004; Wolsink 2000.
[6] Bell et al. 2005; Kahn 2000.
[7] Wolsink 2007:2699.
[8] WGA 2006: p. 6.
[9] Mehos and Owens 2005.
[10] US Department of Energy 2007.
[11] US Department of Energy 2007.

[12] WGA 2006.
[13] 2007, p. 7.
[14] DOE 2007; WGA 2006.
[15] WGA 2006, p. 14.
[16] Stoddard, Abiecunas, and O'Connell, 2006.
[17] WGA 2006, p. 19.
[18] In this regard, the WGA Task Force noted that the BLM has taken steps to facilitate the application and permitting process of solar development on public lands (see Appendix A). Another recommended step is the creation of "solar development zones" (also mentioned in Appendix A).
[19] WGA 2006, p. 21.
[20] WGA 2006.
[21] WGA 2006, pp. 22-23.
[22] WGA 2006, p. 23.
[23] DOE 2007.
[24] CRES News, October 2008, p. 2.
[25] Both the National Park Service and the US Fish and Wildlife Service are under the US Department of the Interior.
[26] http://www.kscland.com/slv_history.
[27] http://www.kmitch.com/Heurfano/sanluis.html.
[28] San Luis Valley Development Resources Group. 2007. Comprehensive Economic Development Strategy.
[29] "For Controversial Wind Farm off Cape Cod, Latest Hurdle is Spiritual." New York Times, January 5, 2010. http://www.nytimes.com/2010/01/05/science/earth/05wind.html.
[30] "Dispute Swirls around Geothermal Project." Klamath Forest Alliance, January 14, 2006. http://klamathforestalliance.org/Newsarticles/newsarticle20060115.html and "Medicine Lake Geothermal Leases Questioned in Appeal." Siskiyou Daily News, March 16, 2010.
[31] The Research Advisory Committee advised that the study should avoid examining the CSP suppliers, their technologies, and factors such as the price of steel—these types of questions were determined to be beyond the scope of the study.
[32] Babiuch and Farhar. 1994.
[33] The study methods were approved by the Human Research Committee at the University of Colorado. Interviewees were promised confidentiality and anonymity.
[34] The energy-water nexus was a major theme of the Natural Resource Law Center's 30th Annual Conference, Western Water Law, Policy, and Management: Ripples, Currents, and New Channels for Inquiry held in June 2009 at the University of Colorado Law School. Sandia National Laboratory has pointed out that there is no national research program directed specifically at understanding the "intimate relationship" between energy and water. http://www.sandia.gov/energy-water/nexus_overview.htm. See also Webber 2008.
[35] This lay description of the SLV hydrology has not been reviewed by water experts.
[36] www.nrel.gov/csp/troughnet/faqs.html. The extent of land needed varies by the amount of thermal energy storage.
[37] For example, beginning in 1964, the town of Parachute, Colorado, suffered though a boom-bust cycle brought on by the Tosco Corporation Colony oil-shale project, the goal of which was to build a commercial scale plant to extract oil from the local shale rocks (Guerrerio 2009).
[38] The 8.22-MW Sun Edison plant was activated on December 17, 2009 near Alamosa, Colorado. It was the largest solar PV plant in the United States supporting substation loads for a major public utility. The plant consists of three distinct types of solar technologies: (1) single-axis tracking array, (2) fixed mount array, and (3) dual axis tracking array with PV concentrator technology. (http://www.sunedison.com; http://www.xcelenergy.com).
[39] The reference is to European CSP companies.
[40] http://www.sdsmartenergy.org/SRPL_mythvsfact.pdf
[41] One reviewer said it could be regulatory rather than statutory authority that is needed.
[42] http://www.ferc.gov/about/about.asp
[43] http://www.nerc.com/page.php?cid=1
[44] http://www.wecc.biz/About/Pages/default.aspx.
[45] http://www.westgov.org/index.php, glossary, accessed April 20, 2010.
[46] http://www.leg.state.co.us.clics.clics2009a/csl.nsf/fsbillcont/DF3A07B67EB2E7FE8725754 7007928B9? Open&file =1300_01.pdf.
[47] For example, the BLM office in the SLV convenes a local citizen group and meets with it regularly.
[48] http://trustlands.state.co.us/Pages/SLB.aspx.

[49] Colorado Governor's Energy Office 2008.
[50] Colorado Governor's Energy Office 2009.
[51] Colorado Governor's Energy Office, 2009. The map of the transmission line referred to is on page 27 of the report.
[52] The requirement applies to utilities serving 40,000 or more customers; and utilities may hold an election of their customers to be included or exempted from the requirement.
[53] Colorado Governor's Energy Office 2009 (REDI Report).
[54] Quoted verbatim from page 7 of the REDI report.
[55] Ibid, p. 8.
[56] Ibid, p. 85.
[57] Ibid, p. 6.
[58] Colorado PUC Combined Docket #08E324 and 325
[59] Ibid., page 15.
[60] Tri-State and Public Service have agreed to an MOU specifying the allocation of ownership, costs, capacity rights, and maintenance agreements for each segment. For example, Public Service will have 60% of the capacity rights in the new San Luis Valley–Calumet segment, and will bear 60% of the costs.
[61] Direct Testimony and Exhibits of Gerald M. Stellern, Xcel Energy. October 28, 2009.
[62] Ibid, page 11.
[63] Rebuttal Testimony and Exhibits, Karen Hyde, Xcel Energy, December 2, 2009.
[64] Ibid. page 7.
[65] The Denver Post.
[66] "Power line project spurs fight over route in southern Colorado," The Gazette, R. Scott Rappold, Nov. 7, 2009.
[67] CBS, Denver, http://cbs4denver.com/business/forbes.ranch.colorado.2.597781.html
[68] Ibid.
[69] Proposed transmission line for San Luis Valley sparks debate," Mark Jaffe, The Denver Post, October 30, 2009.
[70] Answer Testimony of Michael McFadden, On Behalf of Blanca Ranch Holdings, Colorado PUC, December 2, 2009.
[71] Ibid.
[72] Testimony filed to the Colorado PUC by Christine Canaly, Director, The San Luis Valley Ecosystem Council , November 10, 2009.
[73] Areas in the National Landscape Conservation System including National Heritage Areas are governed by other laws requiring protection as a priority to protect objects of historic or scientific interest, and must be managed to protect those values as a priority over other uses.
[74] Ibid.
[75] Ibid.
[76] Surrebuttal Testimony submitted by Tom Darin on Behalf of WRA, January 10, 2010, before the Colorado PUC.
[77] Proposed transmission line for San Luis Valley sparks debate," Mark Jaffe, The Denver Post, October 30, 2009.
[78] Letter submitted Aug., 12, 2008, to Nicole Korbe, Senior Environmental Planner, Re: Tri-State 230-kilovolt Transmission Line Project Proposal (Walsenburg to San Luis Valley Sub-station), by Thomas Spezze, Southwest Region Manager, CDOW
[79] Ibid.
[80] Ibid. The Rio Grande Cutthroat Trout is a candidate species under the Endangered Species Act. And a "Colorado species of special concern."
[81] Ibid.
[82] "Counties support new power line for San Luis Valley," The Denver Post, Mark Jaffe, Nov. 11, 2009.
[83] Ibid.
[84] "DWP drops plan to build 85-mile power transmission line across the desert." 2010. Louis Sahagun, Los Angeles Times, March 11.
[85] Solar Energy Development PEIS Webmaster, solareiswebmaster@anl.gov.
[86] Rebuttal Testimony and Exhibits, Karen Hyde, Xcel Energy, December 2, 2009 (described in Part Two of this report.).
[87] WGA 2006, pp. 22-23.
[88] The Denver Post, December 13, 2009, p. K-8.
[89] A quote from a Valley stakeholder: "There should be a focus group of people to deal with energy issues because there are huge issues today."
[90] Johnson 2010.
[91] Migoya 2010.

[92] Solar Energy Development PEIS Webmaster, solareiswebmaster@anl.gov.

[93] Programmatic EIS Web site or the West-wide Energy Corridor Programmatic EIS, corridoreiswebmaster @anl.gov.

[94] The Interior Department Web site, www.doi.gov/news/09 News Releases/031109c.

[95] The Interior Department Web site, www.doi.gov/news/09_News_Releases/031109c.

[96] Solar Energy Development PEIS Webmaster, solareiswebmaster@anl.gov.

[97] Solar Energy Development PEIS Webmaster, solareiswebmaster@anl.gov.

[98] Solar Energy Development PEIS Webmaster, solareiswebmaster@anl.gov.

[99] Solar Energy Development PEIS Webmaster, solareiswebmaster@anl.gov.

[100] Solar Energy Development PEIS Webmaster at http://solareis.anl.gov/ public comment record no. 60236.

[101] Solar Energy Development PEIS Webmaster at http://solareis.anl.gov/ public comment record no. 60224.

[102] Solar Energy Development PEIS Webmaster at http://solareis.anl.gov/public comment record no. 60126.

[103] Solar Energy Development PEIS Webmaster at http://solareis.anl.gov/ public comment record No. 007.

[104] Solar Energy Development PEIS Webmaster at http://solareis.anl.gov/ public comment record no. 60163.

[105] Solar Energy Development PEIS Webmaster at http://solareis.anl.gov/ public comment record no. 61049.

[106] Solar Energy Development PEIS Webmaster at http://solareis.anl.gov/ public comment record no. 60155.

[107] Solar Energy Development PEIS Webmaster at http://solareis.anl.gov/public comment record no. 60119.

[108] Solar Energy Development PEIS Webmaster at http://solareis.anl.gov/public comment record no. 60064.

[109] Solar Energy Development PEIS Webmaster at http://solareis.anl.gov/public comment record no. 60111.

[110] Solar Energy Development PEIS Webmaster at http://solareis.anl.gov/ public comment record no. 60247.

[111] Solar Energy Development PETS Webmaster at http://solareis.anl.gov/ public comment record No. 61085.

[112] Solar Energy Development PETS Webmaster at http://solareis.anl.gov/public comment record no. 60102.

[113] Solar Energy Development PETS Webmaster at http://solareis.anl.gov/public comment record no. 61049.

[114] Solar Energy Development PEIS Webmaster at http://solareis.anl.gov/public comment record no. 60213.

[115] Bureau of Land Management/Colorado at www.blm.gov/co/st/en/BLM_Programs/energy/ renewable_energy. print.html.

INDEX

D

E

F

G

H

I

J

L

M

N

O

P

Q

R

S

T